ASPECTS OF TOURISM
Series Editors: Chris Cooper *(Oxford Brookes University, UK)*, C. Michael Hall *(University of Canterbury, New Zealand)* and Dallen J. Timothy *(Arizona State University, USA)*

Mallorca and Tourism
History, Economy and Environment

R.J. Buswell

CHANNEL VIEW PUBLICATIONS
Bristol • Buffalo • Toronto

Library of Congress Cataloging in Publication Data
A catalog record for this book is available from the Library of Congress.
Buswell, R.J.
Mallorca and Tourism: History, Economy and Environment/R.J. Buswell.
Aspects of Tourism: 49
Includes bibliographical references and index.
1. Tourism–Spain–Majorca–History. 2. Majorca (Spain)–Economic conditions. 3. Majorca (Spain)–History. 4. Majorca (Spain)–Environmental conditions. I. Title. II. Series.
G155.S6B88 2011
338.4'79146754–dc22
2011015351

British Library Cataloguing in Publication Data
A catalogue entry for this book is available from the British Library.

ISBN-13: 978-1-84541-180-0 (hbk)
ISBN-13: 978-1-84541-179-4 (pbk)

Channel View Publications
UK: St Nicholas House, 31–34 High Street, Bristol BS1 2AW, UK.
USA: UTP, 2250 Military Road, Tonawanda, NY 14150, USA.
Canada: UTP, 5201 Dufferin Street, North York, Ontario M3H 5T8, Canada.

Typeset by Datapage International Ltd.
Printed and bound in Great Britain by Short Run Press Ltd.

Mallorca and Tourism

MIX
Paper from
responsible sources
FSC® C014540

FSC
www.fsc.org

ASPECTS OF TOURISM
Series Editors: **Chris Cooper** *(Oxford Brookes University, UK)*, **C. Michael Hall** *(University of Canterbury, New Zealand)* and **Dallen J. Timothy** *(Arizona State University, USA)*

Aspects of Tourism is an innovative, multifaceted series, which comprises authoritative reference handbooks on global tourism regions, research volumes, texts and monographs. It is designed to provide readers with the latest thinking on tourism worldwide and push back the frontiers of tourism knowledge. The volumes are authoritative, readable and user-friendly, providing accessible sources for further research. Books in the series are commissioned to probe the relationship between tourism and cognate subject areas such as strategy, development, retailing, sport and environmental studies.

Full details of all the books in this series and of all our other publications can be found on http://www.channelviewpublications.com, or by writing to Channel View Publications, St Nicholas House, 31–34 High Street, Bristol BS1 2AW, UK.

Contents

Acknowledgements

Since the beginning of Mallorca's tourism industry in the late 19th century, local academics and those with a professional interest in the island's tourism have produced a rich literature much of which has remained largely hidden from many English readers because it has been written in Catalan and Castilian, the two languages of the island. This book is an attempt to bring to those without the benefit of access to these sources some of the results of the work of Mallorcan scholars. I trust that they will indulge my use of their findings in a reflective way to better illuminate the cause of Mallorcan tourism in order to bring it to a wider audience. Naturally, I have added to these my own research, findings and interpretation.

I have relied heavily on the material available in Mallorcan libraries and consulted over the last 20 years, the period within which I have been visiting the island annually, and often for long periods at a time. No student of any aspect of Mallorcan life can progress without access to the Biblioteca Bartolomé March near Palma's Cathedral. Snr. Fausto Roldán Sierra, its Director and his staff have put at my disposal their very considerable resources. My thanks to them and to Dr Gonçal López Nadal for introducing me to this wonderful library many years ago. The second library that all those with any interest in contemporary tourism in the Balearic Islands should have access to is that of INESTUR (now Agència de Turisme de les Illes Balears), the strategic analysis and information agency for tourism, located at ParcBit near the University of the Balearic Islands (UIB). The librarian Yolanda Pérez Villar and her staff provided me with access to a great deal of material and pleasant working conditions. However, because of governmental reorganisations, the future of the ParcBit building is unfortunately uncertain (see Note in 'Sources and References' chapter). Biblioteca Lluis Alemany in Palma is a major source for secondary historical material. It is hardly surprising that UIB has become a major centre for the study of tourism. Its library at its campus north of Palma has a very comprehensive collection of material. There is also an excellent network of municipal libraries in Mallorca, some of which have provided material on tourism. Three noteworthy ones are those of Manacor, Artà and Palma. The catalogues of most of these libraries are available online. At all of these Mallorcan libraries staff have been most helpful and especially patient given my hesitant spoken language skills.

In the United Kingdom, two libraries and their staff have been particularly helpful: the University of Wales Institute, Cardiff's School of Management Library, which has the best collection of tourism material in South Wales, and the Arts and Social Studies Library of Cardiff University.

Some years ago my first sally into Mallorcan tourism was aided enormously by conversations (in halting French!) with Professor Pere Salvà of the Department of Earth Sciences at UIB. The influence of his early guidance and generosity can be detected in this book. In the geography department of my own university, a trio of academics who know much more about tourism – and especially the tourism of Spain – than I do set an example and a high standard of scholarship that I have struggled to follow. I owe Dr Michael Barke, Dr John Towner and Lesley France a considerable debt. I am obliged to Dr Gonçal López Nadal and Francisco J. Maturana Bis for permission to quote from their paper on the expansion of the Mallorcan hotel industry abroad and to the latter for obtaining for me a copy of *Llibre Blanc* (2009). Teresa Sanchez Cuenca introduced me to the joys of hill walking in the Tramuntana of Mallorca, giving me first-hand insight into one aspect of the diversification of tourism on the island. Karen Hughes and Mark Bellis of Liverpool John Moores University kindly pointed me to their research on the behaviour of young visitors to Mallorca. My friends and neighbours at *urbanització* Palmasol have supported my efforts to come to terms with the benefits and shortfalls of residential second home tourism.

The maps and diagrams have been professionally produced by Alun Rogers of the Department of Earth and Ocean Sciences at Cardiff University. All the photographs come from my own collection. Elinor Robertson and Sarah Williams at Channel View Publications have been my patient editors.

I believe that all the material used in this book falls within the sphere of 'fair dealing' and the usual academic protocols. Should any writer or publisher feel otherwise, I, and the publishers, would be pleased to hear from them.

Lastly, I owe much to the support and encouragement of my partner Hawys Pritchard, herself a noted translator of Spanish, who has shared many visits and much fieldwork in Mallorca. I have often had to neglect her in favour of this book both in Wales and in Mallorca. I trust that she will find the result sufficient recompense.

As in all such works any errors or shortcomings remain mine alone.

R.J. Buswell
Cowbridge and Palmasol, Mallorca
11 July 2010 – a date every Spaniard will remember.

Preface

About This Book

The emphasis in recent research in tourism has been on the processes that have shaped the industry with a longer-term aim of developing theories and constructs that may help explain *development*, defined here simply as change through time and space. This is wholly admirable, but there has been a tendency to drift away from one of the principal origins of the academic approach to tourism that lies in that most eclectic of subjects, geography. Many of the founders of the Academy's study of tourism were geographers, derived from their concern for social and economic activities in different places and in the spatial patterns produced. In this book, I have tried to give back to tourism studies something of the importance of *place* by focussing on the specific characteristics of Mallorca. The primary purpose of this book is to give an account of the development of tourism in the Balearic Islands of Mallorca, the scene of some of the earliest experiments in mass tourism in the Mediterranean.

While wide-ranging it is not entirely comprehensive; as a geographer, for example, I would have liked to include a section on tourism's landscapes in Mallorca but that will be dealt with elsewhere (Buswell, forthcoming). Space considerations also precluded a proper description and analysis of the many social changes tourism has wrought.

It is suggested that only a clear understanding of the *historic processes* that preceded the take-off in the 1950s can reveal the scale of subsequent developments; reaching a state of critical mass was important before the growth curve took on some of the characteristics of the familiar 'S' shape. The focus of this book is *mass tourism*, for which the island is now best known – even notorious –, which had significant antecedents in the late 19th century and the early 1930s. It would be wrong to believe that it sprang fully formed from nowhere. While the book will trace both spatial and temporal patterns in an historic discourse from the 19th century to the present day, examining en route the social, economic and political forces that shaped these patterns, the distinct break-of-slope that occurred in the late 1950s and early 1960s is seen as historically significant.

If the debate about the influence of a long gestation of tourism in Mallorca is one theme, then a second is that the development of tourism has been *transformational*; that is, it has shifted the island's culture from an essentially Catalan/Spanish one towards a much more diverse international and multicultural one. This, it will be argued, is largely explained by processes of inward population migration. The economic needs and wants of the short-term migrants in the form of holidaymakers could not be satisfied by the local economy for reasons of sheer scale, leading to a series of mass population migrations on a more permanent basis from the 1950s onwards and mostly from the poorer regions of Spain. In the last decade, these movements have been followed by apparently permanent migrations from Northern Europe, North Africa and Latin America. Some of the economic and cultural shifts will be examined against this growing and changing demographic background.

Certain elements involved in the transformation of the geography will be examined, emphasising the 'conversion' of Mallorca from an essentially rural island with one dominant city (Palma) into a highly urbanised one consisting of new urban settlements around the coast. This was initially associated with leisure and tourism, the expansion of the primate city into what is increasingly seen as a city region and to a lesser extent, with the development of old inland towns into new centres of population growth providing much improved services to local populations. These are now complemented by a revival in some parts of smaller settlements and isolated dwellings for commuters and new residents relocating from overseas.

The third theme is that of *response*. Whereas tourism has transformed society and economy in Mallorca, the activity is seen as both a success and a problem – a success in that it has created thousands of new jobs and raised per capita income to about the highest level for any part of Spain, and a problem in that it has created a degree of cultural undermining and environmental degradation. Responses to the former have included ways of accommodating affluence, founding and expanding new forms of business including the diversification brought about by the multiplier effect, accepting some form of 'planning' in the local polity and new pressures on educational and training resources. Amongst the responses to the latter, the most dominant has been the development of environmental management, leading later to policies directed towards sustainability.

By the late 19th century, its economy, more advanced than many would have previously imagined, was originally dominated by agriculture and trade with a growing manufacturing sector as it sought to modernise. Its society at the beginning of the 20th century was highly stratified with a small haute bourgeois elite based on the professions, especially of law and medicine, plus the remnants of a landed aristocracy below which was a rural society of small farmers and landless agricultural workers together with a small urban proletariat largely concentrated in Palma. Social class divisions have certainly not disappeared in the subsequent century, but such divisions are much less pronounced, although sometimes exacerbated by seasonal unemployment, a marked characteristic of the tourism industry.

Lastly we ask, are the economic and social responses ascribable only to tourism and its transformational processes or are they in reality little different from social change elsewhere in the developed world, simply formed by a different economic driver – tourism, rather than industrialisation?

The Content of the Book's Chapters

One purpose behind this discourse is to draw attention to the notion of *resources for tourism*, to be dealt with in detail in the subsequent chapters. At one level, for the attraction of tourists to a particular place these are conventionally divided into physical and cultural resources. At another level, they contain elements of such factors as transportation to a destination and within it, hotel and resort structures, entertainment and holiday organisation and management.

Physical resources (Chapter 2) consist of the environmental and geographical characteristics of places – the bays, the beaches and the climate and weather systems. These have proved to be particularly important for Mallorca as it was, and remains, essentially a sun and sand destination despite recent attempts to diversify the product. However, our emphasis will be on the perception and value of these physical resources for the tourist, especially for the tourist *en masse*. The cultural resources might be seen as the features that distract the tourist from the beach, from sunbathing and sea bathing, but, of course, as the structure of holidays changes they have become more than 'distractions' or alternatives within the fortnight's holiday but are resources in their own right, the very things that some types of tourists have come to see (to gaze upon) or participate in. Important here are the cultural infrastructure in an historical sense – often the built environment – and those cultural elements of a contemporary nature – the galleries, the theatres, sport facilities and events, restaurants and gastronomy – some of which, such as outdoor activities, are dependent upon the natural environment. In our analysis of Mallorca's tourism industry, it will be necessary not only to describe these resources but also to account for their geographical distribution over space and time, their consumption by different groups of holidaymakers based on their socio-economic features and the respective role that they play in the industry. Resources by definition and use are dynamic in so far as patterns of supply and demand vary over time and space, by social group and by producers as well as consumers. These themes will be explored in the heart of the book in Chapters 2–5.

A major concern that is always being examined in regions of mass tourism is the degree to which environmental resources are being degraded and so their attractiveness to consumers declines. This is the old adage that 'the tourists destroy the very thing that they seek to enjoy whilst on holiday' and for which they rarely pay. It will be important to distinguish between private and public costs and benefits. Often in mass tourism areas, many of the less visible costs are borne by local taxpayers who need to see a return on their costs to justify the additional social and economic benefits that accrue from having their all-year-round living spaces occupied by visitors for a limited high season. In a geographically bounded space such as the island of Mallorca, this pressure on resources is heightened by the fact that the indigenous population of the island is growing and that a post-touristic economy is emerging, almost independent of tourism. Just because tourism provides about 80% of the island's gross national product, it does not mean that its gross domestic product, which includes its domestic economy outside tourism, is not growing too. It also has its own 'carrying capacity', and it is making increasing demands on island resources that have to be added to those imposed by tourism. An example of this is in the field of housing supply where changing rates of household formation, the immigration of workers and retiring residents require servicing on a not inconsiderable scale, making new demands on construction material supply ('the quarrying of Mallorca') and drawing in imports.

Mallorca's economy and society have become more complex thanks to a lessening of a total dependence upon mass tourism with, now, a broader spread of tourists through the year and an increase in permanent settlers. This is placing new and different pressures on local structures in the public sector such as health, education and social services. It will also be important to give some consideration to the private sector response particularly amongst the owners of hotels – chains and single premises – the tour operators and transport companies. Chapter 8 will try to unravel the symbiotic links between *business* and *politics* within the context of Mallorcan democracy. Chapter 9 concentrates on the *diversification* strategies adopted in Mallorca to try to ease the pressure on resources seasonally and spatially. Many local observers would probably support a diminution in tourist numbers if the income generated by non-seaside activities were sufficient to compensate for the loss. Chapter 10 concentrates on the *proximate future* for tourism in Mallorca, examining the tension between the need to continue with the Balearic Model, which has created the present-day affluence, and the pressures from some quarters to move into more sustainable holiday and leisure markets.

While this book will seek to portray the growth and development of the tourism industry in a particular place, it will be obvious that Mallorca's experience has been the product of international forces – her tourists are exogenous not indigenous. A curious observation is that very few of the analyses of this activity have been contributed by 'outsiders'. Of the millions of words written about this place and its tourism, nearly all have come from the pens and word processors of Balearics, mostly written in Catalan particularly since the 1970s, a language that few students of tourism from elsewhere read or speak. Perhaps this book, with due humility and acknowledgement to Balearic and Spanish scholars, might help fill that void and let a little more light into a world happily experienced by so many outsiders but studied by relatively few of them.

A Note about Languages

Where possible Catalan spelling has been used in preference to Castilian. It is the written language of Mallorca; Mallorquin does have a written form, but it is essentially a spoken dialect of Catalan.

A Note about Statistics

This is not a statistical sourcebook, but it does contain quite a lot of statistics although not a great deal of statistical analysis. Forty years of teaching undergraduates confirms that few of us remember statistics in detail; such numbers often 'dance before the eyes'. In many cases, therefore, I have rounded figures up or down to make them more memorable. It should be noted that in some cases figures quoted are sometimes for the Balearic Islands and not solely for Mallorca.

The sources where detailed statistics of tourism in Mallorca and the Balearic Islands can be found are given in the 'Sources and References' chapter at the end of the book. However, it should be noted that locally produced statistical data on tourism in the Balearic Islands are currently being made to conform to national criteria. This can lead to discrepancies. The statistical data used in this work have been drawn primarily from *El turisme a les Illes Balears: Dades Informatives* (INESTUR, various years), now available online from The Tourism Observatory at www.observatoridelturisme.caib.es.

Figure 1 Map of coastal resorts

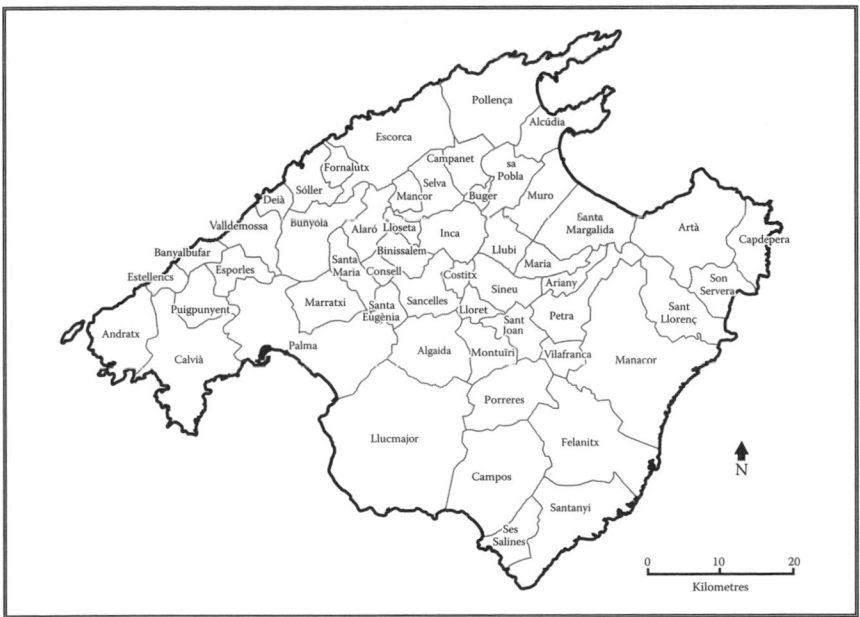

Figure 2 Map of municipalities

Chapter 1
An Introduction to Mallorca's Tourism and its Origins

Links between Past and Present?

To many Mallorcan writers on tourism there seems to be a strong connection between the perceptions of early visitors and the development of the modern industry, but, as we argued in the Preface, this link over such a long period of time is somewhat tenuous. In Chapter 4, we shall show that the development of mass tourism in Mallorca in the 1950s was a 'marked change of slope', a 'paradigm shift' in the history of tourism, a 'break with the past', not an evolutionary movement but a step-like change, the result of a government policy deliberately seeking non-tourist macroeconomic objectives tied to the development of the state. This is a somewhat different process from that described by writers such as Walton in their analysis of tourism and holidaymaking in Great Britain, although Barton points to certain similarities (Barton, 2005; Walton, 2002, 2009; Walton & Walvin, 1983). Mass tourism in Mallorca exhibits rather different characteristics from the historical development of seaside holidays in Britain's Blackpool or on Germany's Baltic coast. Nonetheless, early tourism in Mallorca and the relaying of its experiences did form some foundation for later growth. Thomas Cook may have organised tours to Mallorca and cruise liners in the 1930s may have called in at Palma, but there is little evidence from the literature that the British pioneers of *mass* tourism to the island were aware of, or influenced by, earlier visitors. Barton, however, provides a valuable insight into the role of 'sending' organisations in the 1930s and 1950s before the first wave of mass tourism took off in the late 1950s (Barton, 2004, 2005). Similarly, John Walton has given us a useful study of tourism in one corner of Mallorca in this early period (Walton, 2005).

However, the processes involved in the historical development of tourism are complex and recently Mallorcan researchers have given considerably more attention to what took place in Mallorca itself before the onset of mass tourism whereas most British writers see its take-off from the mid-1950s. Local researchers point in particular to the significance of domestic Spanish tourists in the 1920s and 1930s, a factor identified by Barke and Towner for mainland Spain and by others, notably Cirer, for Mallorca (Barke & Towner, 1996: 26; Caro Mesquida, 2002; Cirer, 2006, 2009). Chapter 3 will examine this in more detail.

Large-scale tourism – the prime concern of this book – may have its roots partly in pre–Civil War Mallorca itself, but it is our contention that it is mostly a feature of post–Second World War Europe, partly the product of Mallorcan opportunism and partly the product of British – and later German – entrepreneurship. Nonetheless, it is important to have some insight into early visits and visitors to Mallorca before tourism became widely organised commercially.

The Perceptions of Some 19th-Century Visitors from Northern Europe to Mallorca before about 1880

For many of those who initially visited the island as travellers it was the nature of Mallorca's microcosmic world that was a major attraction: here were plains, hills, vertiginous mountains, beaches and coves, forests and *garriga* (typical low-growing Mediterranean vegetation – *maquis* in French), a rural aristocracy and a landless peasantry in curious costumes – all within a day's journey from a Palma hotel. Added to this was the Mediterranean climate, especially in the winter when it had an added attraction for North Europeans, presenting the possibility of warm sunshine when much of the continent was snowbound. Further enquiry revealed a rich cultural heritage ranging from late Bronze Age stone structures – the *talayots* – to the fine medieval and renaissance buildings of a considerable city, Palma. In the countryside was a class of landed gentry and their improving estates – the *possessions* – some with formal gardens, most producing exotic tree crops such as figs, carobs and almonds, others vines and olive oil. What Mallorca and the other islands lacked historically, of course, was a proper link to Classical Antiquity, which, alongside its island location, was the major reason it was not on the itinerary of the Grand Tourists of the 18th and early 19th centuries. The island was invaded and settled by the Romans in the first-century BC but left few remains to attract the classicists, with the exception of the Roman town of Pollentia in the north, even now not fully revealed. Even if, as Barke and Towner explain, Spain may have had a burgeoning domestic leisure industry as early as the 18th century, Mallorca had few of the other 'attractions' that young English gentlemen required from such destinations as part of their broadening education (Barke & Towner, 1996: 22; Towner, 1996).

Amongst the early visitors – and even today the most famous – were George Sand (Madame Dudevant) and Frederick Chopin who came for health reasons in 1837. Sand initially found the island attractive, but she ended by loathing it. Her later accounts could surely have done little to attract future tourists (Sand, 1855).

British travellers in the 19th century observed the lack of facilities such as decent hotels for tourists. As early as 1809, Sir John Carr observed that

Palma's inns were 'very few and very bad' though he himself was well accommodated (Carr, 1811: 331). Captain Clayton complained bitterly of his fonda, *Tres Palomas* (Clayton, 1869), and the American writer Bayard Taylor was cordially received at his fonda, the Four Nations, but '... afterwards roundly swindled' (Taylor, 1867–1868: 680). Charles Toll Bidwell noted: '... if Palma had a large and well-organized hotel ... its attractions would be increased, and more visitors would lighten it with their countenance ... Fifty years ago such a thing as an inn was said to be unknown in Majorca', although he points to recent improvements (Bidwell, 1876: 54). Many of these early visitors remarked upon the rather shallow social life and the unhealthiness of many parts of the island, with outbreaks of cholera and malaria occurring frequently. In addition, when faced with an often difficult sea voyage from Barcelona or Marseilles, it was hardly surprising that few north Europeans made Mallorca a destination.

Those who did visit found the island physically attractive, especially the mountains. E.G. Bartholomew, for example, a British engineer engaged in laying the first undersea telegraph linking the island to the mainland in the 1860s, called the scenery 'magnificently varied. The mountains bold in the extreme' (Bartholomew, 1869: 266). Captain Clayton, also writing in the 1860s, wrote rather extravagantly of Puig Major, the highest point in the island, 'Raised high above a region of mountain peaks, black stupendous gorges, and a wild chaos of riven rocks, shot up from the bowels of the globe in some primeval convulsion, soars the massive summit' (Clayton, 1869: 238). The ordinary people – the 'natives' – although described as content and friendly were often seen by the class-conscious British as insubordinate or over-familiar while the Mallorcan aristocracy and gentry were thought to be lazy and indolent. There is an almost anthropological thread running through many of these early narratives. For example, Charles Toll Bidwell, the British consul in the 1870s, wrote of the small farmers in the Albufera area:

> The faces of the men were peculiar. A few were decent, but many were repulsive. After all what can be expected of them? What else can be the result of lowly birth, coarse surroundings, hard lives and scanty fare? Here a man of fifty is wrinkled and curved and looks eighty; a man of seventy might well have come out of Noah's ark. (Bidwell, 1876)

This kind of description is perhaps not so surprising given the imperial or colonial background of many British and French visitors. These early visits were, of course, largely before the beach and its sea bathing became an important resource for tourism and so there are few accounts of the seashore. What was more important was the cultural landscape: the churches, monasteries (before 1838), the city of Palma, the country houses and estates of the gentry. In any case, that is what so much of early tourism was about: the search for the exotic, the different, the 'other'. It was believed that there was no 'lesson' to be learned from Mallorca as there was from the remnants of classical Italy or Greece for the 18th-century Grand Tourists. The language – Mallorquin, a dialect of Catalan – added to the exoticism as did the ubiquity of the Catholic religion about which many Protestant visitors remained cautiously curious. Bidwell noted the peculiarity of street processions at the time of various fiestas. Many of the British gave the impression that the Inquisition was never far away. Perhaps these were also amongst the reasons why so few foreign visitors came to Mallorca. In 1875, Bidwell recorded '39 British, 119 French, 43 Italians, 13 Americans and 1 Swede' (Bidwell 1876: 20). American visitors were more frequent than might be expected especially those of a literary bent such as Bayard Taylor, a friend of Ralph Waldo Emerson and Mark Twain, whose account is rather more positive about Mallorca than most (Taylor, 1867–1868).

Some outsiders took a much more academic view of the island, describing and analysing its make-up in more scientific detail. The doyen of these was Archduke Ludwig Salvator (1847–1915), Ludwig Salvator d'Hapsburg-Lorena, ninth son of the Grand Duke of Tuscany, Leopold II – a descendant of the Austro-Hungarian emperors – and Maria Antonieta; he first came to Mallorca in 1867. He has become something of a legendary figure on the island partly because of his alleged prowess with local women but more important for us as a major influence on part of the island's western landscape. On his father's side, he was cousin of Franz Josef of Austria, assassinated at Sarajevo. He was born in the Pitti Palace in Florence, studied philosophy and natural sciences in Vienna and Prague and spoke at least nine languages, including Catalan and Castilian. He eventually owned land and large houses in Zindis near Trieste, Ranleh near Alexandria in Egypt and Nice as well as acquiring large areas in Mallorca. From the 1870s he built up a formidable estate in the north and west of the island that he proceeded, in part, to landscape mostly on romantic lines rather than use it as a source of agricultural wealth. He was fascinated by all things Mallorcan and even went as far as acquiring a Mallorcan mistress. However, his great literary work was *Die Balearen*, published in nine volumes in Leipzig from 1881. This gave detailed accounts of all aspects of the island: geography, history, economy, biology and anthropology. Although eventually translated into Castilian, regrettably, it has never been translated into English. How influential it was in attracting other German-speaking visitors to the island is doubtful. His legacy today lies more in his lands in and around the mountains that have helped form the basis for much of the hill walking now of increasing popularity in low-season Mallorca (Cañellas Serrano, 1997).

Navel and military personnel, especially French and British (e.g. see Carr, 1811), were another group of early visitors. Their presence was largely the result of competition between them for control of the western Mediterranean. Much of this interest declined by the mid-19th century, but the strategic location of Mallorca again became of interest to British and German concerns with the success of Franco in the Spanish Civil War and the possibility of Spain joining the Axis forces. But spies probably cannot be counted as tourists.

An Island in the Mediterranean

It is, in any case, very difficult to be sure that these early visitors and their perceptions of Mallorca were to influence the flow of commercial tourists in the late 19th century and after, but clearly the fact that Mallorca was an island gave the place a certain intrinsic fascination, even intimacy (Trauer & Ryan, 2005). Some have seen the Balearics and indeed many other Mediterranean islands as stepping stones – an idea promoted earlier by Braudel (1992: 116) – that they were places on the way to somewhere else, but in the case of Mallorca its history suggests that it had a distinct and separate identity and culture. It was not simply 'on the way to somewhere'; it was 'somewhere' in its own right. This simple idea will become important in our consideration of the development of the Mallorcan economy and the role of tourism within it. Mallorca was not a place apart nor isolated from wider European cultural events because it was a small dot in an ocean. Historically, it developed rich cultures of its own based partly on this very connectedness. For a brief period Mallorca had actually been a kingdom with continental possessions – more of a launching pad than a stepping stone. It is not possible here to go into the origins of this cultural distinctiveness, although we shall return to it when considering aspects of modern development of cultural tourism. However, by the 20th century, the enterprise of Mallorcan businesses that developed mass tourism was in large part based on their economic historical experience dating from the late 18th century and their knowledge of the workings of the wider Spanish state under various leaders from Mendelez to Franco and later still of an even wider European economy under a democratic system that was eventually to make Mallorca one of the richest places in the country and the European Community.

If there is a set of internal, place-specific, factors that help us to understand the early development of the tourism industry in the Balearic Islands – and in Mallorca in particular – it is intimately bound up with their Mediterranean context spatially, environmentally and temporally. The climate, the vegetation, geology and landforms – their physical landscapes – are at once easily identified with the classical characteristics of this 'middle sea'. Indeed, islands are an important part of any definition of what is Mediterranean. As we shall see, 'Mediterranean' can have many meanings to different social and political groups but in most of them the sea and the interconnectedness of places is central. If the sea is a 'constant', then the various landfalls must include the coastal perimeter and the intervening islands. Indeed, islands in the Mediterranean, since most of them are small, are perhaps more typical of mediterraneanism than the continental coastlands because at least one definition of 'Mediterranean' is the climatic one, which, of course, extends inland, often formerly defined by certain distributions of vegetation, most noticeably of the olive. It is this physical Mediterranean that formed such an important resource base for the tourism industry and so an analysis is given in a subsequent chapter. In the post-modern world of mass tourism in the late 20th and early 21st centuries, most environmental influences are relegated below ubiquitous cultural ones.

But the Mediterranean is more than a physical entity or set of physical characteristics; it is also an historical concept, which some date from the Romans' notion of it being their sea or from their viewpoint, *mare nostrum*, 'our sea'. This led to the idea that if there was some kind of geographical hegemony asserted by the Romans through their Empire, then a unity of culture might be possible across it. This, in turn, led (eventually) to Braudel's well-known historical work that was – as was the case of so much 20th- century French history – influenced by geography, first of a deterministic physical kind and later by the more possibilistic ideas of Vidal de la Blache. It is thought that the idea of a physical unity to the Mediterranean may well have been shaped by the Renaissance painters and the effect of an education based on the classics. Later in British Public Schools, at least, the subject of geography was really an adjunct to the history of the classical period; it was where history happened, and the 'places' of that history had to be learned and construed as much as the classic texts. The interconnectedness of places that persuaded Braudel to see a unity of experience and culture in the Mediterranean brought about by ship technology, trade and notions of Empire might also have been influenced by the fact that he spent some of his early life as a lycee teacher in Algeria, part of the French Empire, across this sea. His view of islands, that they were stepping stones between the more important centres located on the various mainlands, is a view more recently reiterated by David Abulafia who saw Mallorca as a hub at the centre of many trade routes (Abulafia, 2002: 48).

The geographical location of the Balearic Islands, whilst a mundane fact in itself, is very important to their development as a tourist destination. The Mediterranean has until recently always been seen as a European sea and the fact that the islands are located 180 km from the north African coast is much less relevant than the fact that they are to be found 120 km from the coast of Spain at Barcelona, 90 km from Valencia and 180 km from Marseilles. The flows of tourists were, and are, overwhelmingly north to south, from the European continent to the islands. Historically, this perception of relative place has much less validity. If the islands were stepping stones, they acted as such between Christian Europe and Muslim Africa where trade flourished in both directions for nearly 2000 years and later between various colonial possessions in North Africa (Morocco, Algeria, Libya, Egypt etc.) and amongst the islands of the Mediterranean (Malta, Cyprus etc.). In the 1920s and 1930s, the islands may have acted as ports of call for the cruise ship industry, but in the post-war modern, mass tourism industry, Mallorca and the other islands are, of course, not stepping stones at all but final destinations: termini.

The fact that Mallorca is an island in the Mediterranean is important for tourism for two reasons: firstly, it involved travel to it, initially by sea and later overwhelmingly by air. This idea of movement over the sea was an important component of the tourist experience. Many early writers spoke of the romance of arriving in Palma harbour at dawn from Barcelona by sea and seeing the Gothic Cathedral bathed in the morning light, itself an unexpected cultural experience (Boyd, 1911: 4; Wood, 1888: 20). Later, of course, travel by air had none of this romance, in fact, quite the opposite. Nonetheless, the notion and experience of travel over water from continent to island had a curious attraction of its own. This was the beginning of a second factor, which this exoticism was to continue for visitors once on the island. Mallorca was perceived to be a 'place apart', a microcosmic world where a different life could be experienced by the visitor if only for the winter season or later for a brief two-week vacation, where perhaps the normal rules of behaviour did not apply, a perception built upon, manufactured by the early travel writers, like the subject of Lawrence's story: 'He wanted an island all of his own; not necessarily to be alone on it, but to make it a world of his own ... a minute world of pure perfection, made by man himself' (Lawrence, 1955); a theme taken up later by the tour operators in their brochures (Baum, 1998: 117–120). By the 1920s, Mallorca was being billed as the honeymoon island, a fabrication of marketing departments. In her novel *Crewe Train* (1926), Rose Macauly has her protagonists honeymooning in the Balearic Islands: 'The exquisite lovliness of Mallorca, its gentle airs and mild winter suns, its great sweet oranges, little deep-streeted Moorish towns, ancient inns and blue bays, its olive-grown mountains, straying, jingling goats and beautiful and amenable inhabitants ...' Mulet records 6000 honeymooners, also entirely Spanish, coming to Mallorca in 1945 (Mulet, 1945: 39).

Many have tried to account for the fascination of islands: isolation, separateness, smallness – giving easy accessibility to the whole and their microcosmic nature as some of the common factors (King, 1993: 13). Although Mallorca and its city were a major centre of trade in the medieval and early modern periods – Mallorcan trade ships were in London and Antwerp as early as the late 13th century – knowledge and perception of the Balearic Islands by the 18th century was primarily fashioned by their strategic position in the western Mediterranean basin. Not only were they on the 'hinge' between Christian Europe and Muslim North Africa, more importantly they offered the potential for a base for the British and other navies in their role of protecting the route to the East and especially India. This imperial role was confirmed by various British occupations of Menorca between 1708 and1802. Mallorca was considered for conquest too in order to deny it to the French, a consideration repeated in the Peninsular Wars of the early 19th century. Thus, much of the early writing and cartography about Mallorca and the rest of the archipelago are of military and imperial origin, if not actually colonial.

Changing Perceptions: Mallorca Then and Now

The perceptions of early visitors that contributed to their fascination with Mallorca was in part, and perhaps largely, an historical and geographical fiction, a product of a manufactured image. It was not until the early 20th century that north Europeans settled permanently in Mallorca in any number and so their view of the island was, until then, principally the result of fleeting visits and short stays and reports home. We shall see in the next chapter that the growth of various artists' colonies in Mallorca led to a more intelligent and deeper understanding of the island's history and culture.

Before the 19th century and for long after, and in reality, there was a sharp division between *Ciutat*, the city, and *part forana*, the remainder of the island. To the visitors before the mid-19th century, the mountains and much of the coast were difficult to access despite their obvious attraction thanks largely to an absence of decent roads. The beaches and bays – and the coast generally – were deserted, often a malaria-infested wilderness likely to be attacked by corsairs and pirates in search of booty and slaves. The country estates were often neglected, their owners deep-in-debt absentees. The city's Cathedral was – and remains – incomplete and its other architectural heritage was poorly maintained with the aristocracy's town mansions surrounded by a sea of mean streets whose pattern dated from the Muslim era. Social divisions were most marked. This does not mean that the island was in some way backward economically; indeed, considerable riches were earned from trade, from olive oil and a thriving textile industry, often its per capita income being above the Spanish average. There are thus at least two 'histories' to be considered: that of the potential visitor, traveller or tourist and that of the Mallorcans themselves, what Jacqueline Waldren has dubbed 'insiders and outsiders' (Waldren, 1996). The latter group also has to embrace the 'Spanish', that is the Castilian-speaking peoples from the Peninsula who have incorporated the island into what was to become 'Spain' since the conquest of the Muslims in 1229. To the Spanish Crown, Mallorca was seen as a peripheral part of its state, geographically on the frontier between its empire and those of the 'moors', the Turks, the French, the British, the Bourbons and so forth, sometimes a staging post to further conquests in the Mediterranean, at other times a place of retreat. In the mid-16th century, Phillip II even contemplated evacuating Menorca's population to Mallorca, seeing the smaller island as indefensible.

Neither of these two scenarios is either entirely correct or incorrect. These simplistic descriptions are merely devices to enable the reader to appreciate the intricacies of the role of perception on decision making with regard to potential visits to places, something that will be discussed at greater length in later chapters. But the historically more objective picture was not the one seen or much known about or understood by visitors or tourists. The picture that they had was the one painted above initially. And that remains one of the major differences between the 'local' and the 'visitor', between the knowledgeable and inquisitive and the superficial concern of the leisured outsider. The tourist's gaze was at odds with reality but if it had not been then the propensity to visit the island at all would probably not have existed. Of course, this difference between the day-to-day living reality of Mallorca and what the tourist now perceives as the reality remains; probably, in this age of mass tourism it is even more pronounced; some visitors are largely unaware that Mallorca is an island, even fewer that it is part of Spain. But when have visitors or tourists ever really wanted to know the 'truth' about where they are staying? Their consumption is of a deliberately created alternative reality; it is this that is the actual attraction. Something too much like home would have been inappropriate, but some kind of simulacrum would give comfort and assurance. Perceptions of 'needs' have changed markedly over time from those of the well-heeled traveller of the early 20th century who required luxury from their hotel, to the first experience of Mediterranean food of the workers' tours of the early post-war period, to the mirroring of the English or German 'High Street' that typifies so many Mallorcan resorts today. There had to be something to be escaped from for a brief period but also something of the familiar.

Many observers of the history of tourism have seen holidays as something akin to seasonal migration and have used models of population movement to explain development. If these perceptions and 'facts' about Mallorca might be seen as 'pull' factors, then it is important to have some understanding of 'push' factors too. Indeed, a better explanation for the early success of the island's mass tourism industry might be drawn from a study of 'demand' rather than 'supply'. Among the factors to be examined here are the development of the notion, firstly, of a leisured class amongst the bourgeoisie, and later, of 'holidays' linked to reduced working hours, the increase in leisure time for the industrial and clerical workers with their rising disposable incomes. However, there is a wealth of difference between the traditional history of holidays on the Atlantic, Baltic and North Sea coasts taken by north European workers in their respective countries and the mass seasonal migrations to the Mediterranean from the mid-1950s onwards and so the traditional explanations of writers such as Walton and Urry may not be easily transferable. What many Mallorcan analysts have turned to instead is the Balearic Model: cheap flights, well-equipped but low-cost three-star hotels, easy access to a clean beach and safe sea bathing, an accessible, familiar social environment, all the product of the close cooperation between local hoteliers and British or German travel companies and tour operators who have combined their skills and resources to provide the ubiquitous package holiday. It was a system of tourist provision that met both supply and demand requirements almost perfectly (Middleton, 1991; Morgan, 1991). Founded in the late 1950s and continuing with little alteration to at least the 1980s, this model laid the foundation for Mallorca's fortunes and was widely copied with local variations in most Mediterranean countries. Despite efforts over the subsequent 60 years, large elements of this model persist today and are so embedded that it is difficult to see their replacement in the proximate future. However, it is important to be able to demonstrate that this model is not entirely inflexible and incapable of change. Pressure from environmentalists in particular has led to considerable legislation to constrain coastal development and to manage resources more carefully and has led, in turn, to hoteliers and tour operators reforming many of their practices not only in response to changing consumer demands but also for internal economic reasons associated with falling margins from previous eras of investment. There has also been considerable effort invested in diversifying the tourism product and changing the seasonal balance. Many different forms of leisure activity now take place on the island, but the hegemony of the five 'Ss' (sea, sand, sun, sangria and sex) remains a very powerful form of holidaymaking for many of the 9 million visitors who come to Mallorca each year.

Chapter 2

Environmental Resources, Perceptions and Constraints for Tourism

Resources: What Do They Mean?

In Mallorca, as in much of the Mediterranean Basin, the most important resources for the development of tourism have been its climate, its surface (physical) features, particularly its coast and beaches and its mountains, and to a much lesser extent its vegetation. To this trilogy should be added the totality of the landscape, a product of the interaction of human society with these physical characteristics over the last 4000 years (Buswell, forthcoming). It is already clear from Chapter 1's comments on the 'fascination of islands' that visitors to Mallorca perceived and used these elements in quite different ways at different historical periods and so to those who came in the second half of the 19th century it was the romanticism of the mountains of Tramuntana that held most appeal together with the mystery of the cave systems. The painters, poets and writers who visited and settled in the island in the period up to about 1930 seemed to favour the landscape as their resource. Although sea bathing had been introduced as early as the 1870s, it was not until the 'Americanisation' of sun culture in the 1920s and 1930s that the beaches became the focus of tourists' attention, a resource base that has remained dominant until today. Of course, all through these historical periods, the Mediterranean climate has been perhaps a constant environmental attraction: in the late 19th century, it was the winter climate, but by the second half of the next century, it was the strength of the summer sun and the blue skies of the Azores and Sahara high-pressure systems.

The conventional wisdom in many texts on the relationship between tourism and the physical environment is to see it as somehow deterministic, that is, that the environment presents obvious attractions for the holidaymaker: if there were beaches they would be lain upon and swam from; if there were mountains they would be climbed. A second view is rather more possibilistic, that is, the environment provides opportunities that certain socio-economic conditions would assist in its exploitation. A more modern behavioural view is derived from perceptual psychology with culture being the filter through which potential decisions about the use of environmental resources are taken. 'Culture' in this sense has many

11

components, but economic status, social class, gender and ethnicity all figure strongly. Yet another view is to recognise a time-dependent dynamism about physical resources for tourism; different eras will have perceived the potential of the physical resources in quite different ways. Concepts of leisure or non-work time itself are a product of certain cultural norms or aspirations or practices. Lastly, it is obvious that resources are not ubiquitous in their spatial distribution but sporadic, punctiform or patchy.

It is possible, then, to construct a set of dimensions to this problem. On one axis will be the cultural characteristics of the resource consumers (the tourists) and on the other a temporal element. Added to this is the notion that resource use is not consistent through time and space; that is, there is a developmental element to any model of resources and tourism. Although the resources may exist at many points in space, they will not all be developed at the same time. There may be patterns of geographical diffusion that can be traced as the phenomenon of tourism spreads, or does not spread, from one place to another. In this sense, the geographical dynamics and the resulting patterns of tourism are not really very different from other forms of social and economic activity.

It is also true that the very resources themselves have by no means remained constant. On the one hand, the beaches and the beach–dune system of which most of them are part have been seriously degraded by human use, with the actual shoreline being shifted by urbanisation, by promenade construction and by the building of marinas with patterns of erosion and deposition consequently being changed. The mountain environments, which are becoming increasingly popular for low-season tourism, are also being eroded by walker and hiker activities in the most popular and accessible locations. On a longer time scale, the whole process of geological relationships is constantly changing as natural processes contribute to land surface movements. The often sudden and step-like nature of geomorphological change has little effect on most tourists, but subtle alterations to the landscape do occur. Examples might include the seismic shifts characteristic of the Mediterranean Basin where the African and European tectonic plates meet or on a much more local and frequent scale the explosive floods and inundations and associated erosive effects that often result from sudden autumnal downpours.

Climate and Weather

Perhaps the major attraction for early tourists to Mallorca was its winter climate, perceived to be warmer and drier than its north European counterpart, with benefits to health (Figure 2.1). Clearly so thought George Sand when she persuaded Chopin to move there in 1838. The disastrous outcome of their visit was in part attributable to their misunderstanding

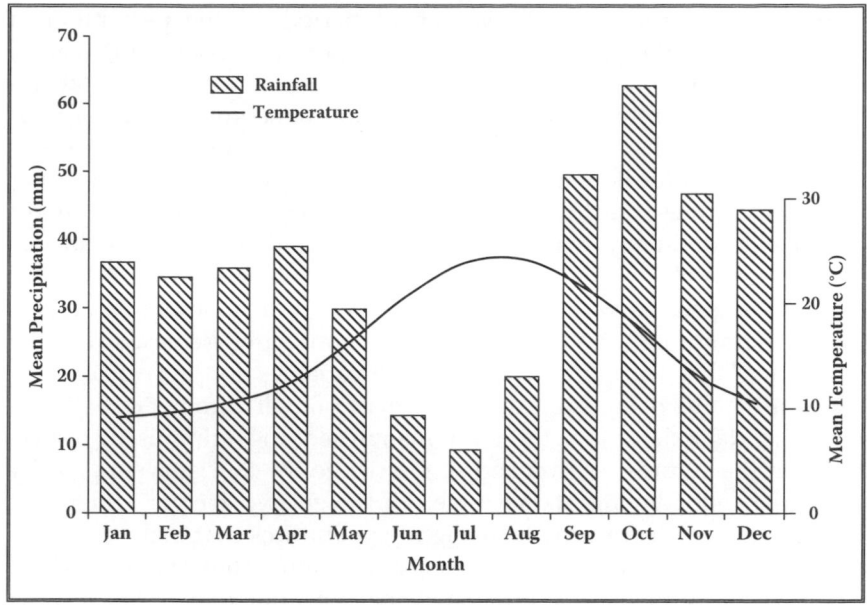

Figure 2.1 Graph of rainfall and temperature distribution

of west Mediterranean winters; Sand writes of 'lugubrious rain'. Bidwell writing in the 1870s thought that the winter climate was not suitable for the invalid:

> ... native consumptive patients go off very suddenly and quickly ... The changes of temperature which take place in the winter months are very sudden, as, in a season which one is tempted to believe is perpetual spring, a storm sets in, often accompanied by snow and hail ... and then many of the brightest days of winter and spring are those which make ladies go out with parasols and muffs at the same time. (Bidwell, 1876: 12–14)

When Gertrude Stein suggested to Robert Graves that Mallorca might make a suitable home for an impoverished writer recovering from the First World War, her recommendation was based more on a low cost of living than detailed knowledge of island meteorology. Graves himself explaining why he chose Mallorca in 1929 wrote in 1953 that Mallorca's climate 'had the reputation of being better than any other in Europe'. But early visitors naturally avoided the heat of high summer. The Victorian and Edwardians thought the pale complexion was the desired appearance; the suntan smacked of the oriental, the native, the 'other'. As we have already hinted, it was largely an American notion that sunbathing became fashionable in the 1920s.

What is important for tourists of all periods is the perception of climate and weather in resorts and if those perceptions (and the reality) do not coincide, whether there are compensations for climatic 'aberations'. Tourists rarely consult the climatic and meteorological record but rely upon generalisation broadcast by travel agents and their manufactured 'image'. Information given about climate is of a very general nature and concentrates on the 'long run'. The meteorological picture of Mallorca is definitely one based on climate and not weather and so high season is always described as hot and dry, winters as warm and the two mid seasons as variable. In truth, there is much variability around these generalisations, a variability that has noticeably increased with the onset of global warming. The two key variables for the vacationing tourist are hours of sunshine and amount of rainfall (Sumner *et al.*, 1993). Long-run data show that these are maximised and minimised, respectively, in the summer season and so the risk to the tourist is supposedly reduced by holidaying at that time. However, two factors undermine this to a certain degree: savagely high temperatures that make exposure to too much sunshine hazardous and thunderous rainstorms of almost tropical intensity. While the former can sometimes last for some weeks, by definition the latter are more sporadic. The combined effect if it coincides with a person's holiday can be disastrous, hence the need, on the part of resort and hotel managers, to offer compensations. The widespread use of air conditioning in hotels, restaurants and bars has reduced the impact of one of these variables.

Climate remains one of the most important resources for all Mediterranean resorts, nonetheless. But it, like aspects of the physical environment, is no longer a determining feature. Holidaymakers to Mallorca seek to balance their meagre knowledge of the island's climate with their expectations of success in weather terms. Their risk calculations may be poor, but their increasing knowledge of the supply of compensating provisions reduces their likelihood of disappointment, especially as outdoor activities give way to indoor ones and daytime activities are compensated by the 24-hour resort, particularly on the part of young holidaymakers (see Chapter 9).

One aspect of climate and weather that dominates much of the current reform movement in Mallorcan tourism is seasonality (Figure 2.2). As the island's planners seek to increase tourist activity and numbers in the seasons outside the summer peak, one difficulty that they are faced with is the variability of the weather during spring and autumn. A characteristic of the increasing popularity of off-season holidays is its attraction for older tourists and those with fewer family ties and ties to traditional non-work periods. These social groups are perhaps less constrained by the need for sunshine, swimming and the beach. A similar rational applies to the semi-resident tourist whose house or apartment offers the same

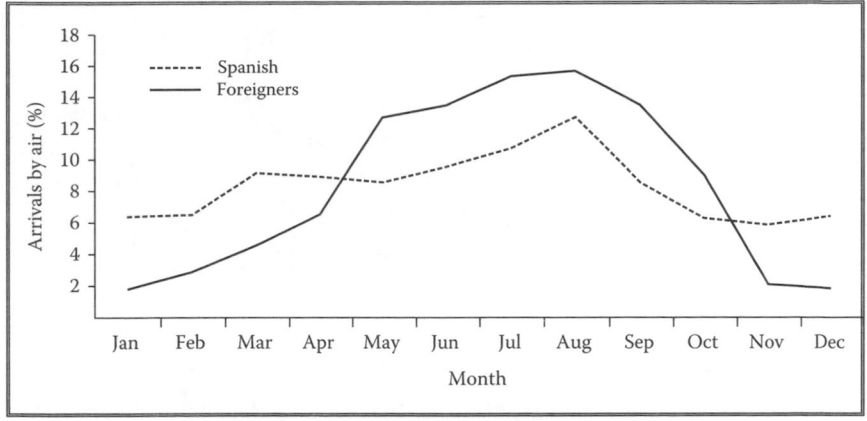

Figure 2.2 Graph of seasonality of visitor flows
Source: *Dades Informatives* (2008)

kind of refuge from more disagreeable or extreme weather as it would at home.

Landforms

In Mallorca, three principal landform elements have played a major part in the development of tourism: the coast, the mountains and the plain (Figure 2.3).

The coast and the beach

Mallorca has a coastline of some 555 km including 36 small islets and two sizeable islands, Cabrera and Dragonera. The total length of the beach in the Balearic Islands is approximately 100 km or about 13% of the coast. These sandy and pebbly stretches are the resource for sunbathing and swimming amongst other activities, core values for the Balearic type of mass tourism. Of the 139 recognised beaches in Mallorca, only a small minority have escaped development usually because they are inaccessible although, of course, most beaches started their tourist lives being so described; relative isolation was a major attraction for many early tourists. Some, such as Cala Mondragó and Es Trenc, have been kept away from the clutches of developers by hard-won and often late legislation, their management being taken into the public domain. Others have remained in private hands, conservative landowners who abjure tourism. Given the relatively small extent of beaches in Mallorca (area and length) and the fact that the island has about 9 million visitors annually means that the densities of potential occupation are very high but spatial variations are considerable, with some municipalities having

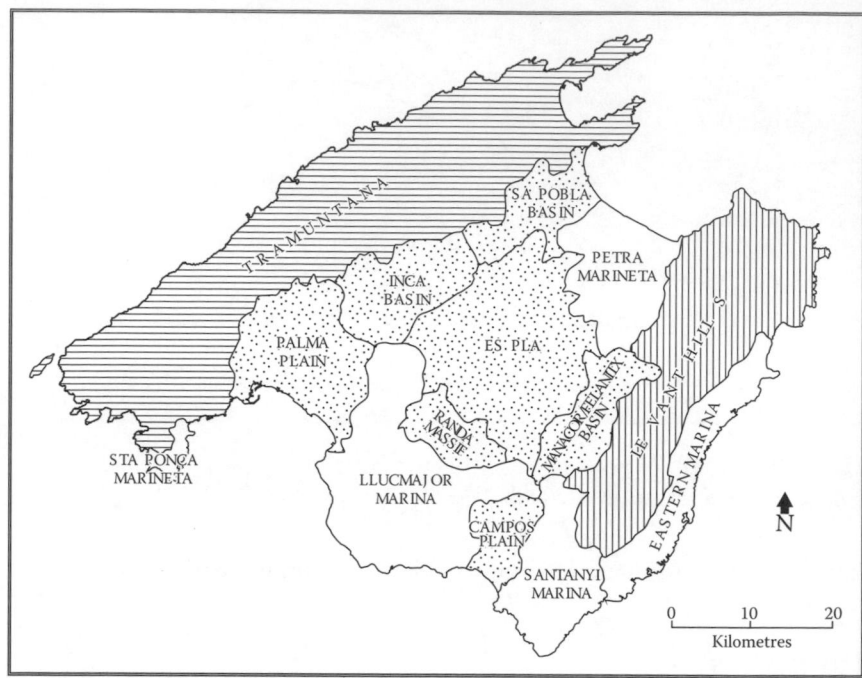

Figure 2.3 Map of physical regions
Source: Atles de les illes Balears (1995)

about the space of a beach towel per visitor in the high season. In Calvià, Andratx, Palma, Llucmajor, Felanitx and Manacor, where for the most part small beaches enclosed by headlands (*cales*) predominate, the pressure on beach space is excessive, leading to problems of physical and, often, social erosion (Buswell, 1996: Fig. 12.2, 322).

The remainder of the coastline is perceived and used in various ways, including for ports, harbours and marinas and as building lands for hotels, apartments and resorts, particularly on the rocky headlands that enclose beaches on the east coast of Mallorca. The steep cliffs of the north and north-west coasts that plunge almost vertically into the sea are accessible only by sea or to the skilled hill walker; their main asset to tourism is their scenic value. Since tourism in Mallorca has manifested itself primarily as an urban function, the suitability of the coastal features for building purposes has proved to be important. Here, Mallorca's soft sedimentary geology has enabled the ready construction of tall hotel and resort complexes on an easily excavated set of bases. Few of the tourist coastal urbanisations extend more than 2–3 km inland. However, it is the beach that is the most important morphological attraction for tourism, but it cannot be detached from its surrounding solid geological structures.

The beach

In the case of Mallorca, there is a misplaced sense that the seaside – the beach – and the nature of the Mediterranean climate are the only significant environmental elements that need to be considered, but both of these elements have played different roles at different times, in different places. Most early visitors were from middle-class backgrounds whether from Spain or other European countries, showing little concern for the beach environment; their major interest – their gaze – was in other aspects of the physical environment, especially the mountains, and cultural artefacts and events. It is also true that the attractiveness of the climate for visitors was in the winter months when a warmer climate than that of northern Europe and the clear skies were thought to be beneficial; the glaring sun and the searing temperatures of July and August were to be avoided. By the 1930s, the cult of the tan had reached Mallorca, probably derived from North Americans' use of coastal Florida and Georgia. The benefits of sunshine for health reasons, especially against the Victorian scourge of rickets, were known by then. The Americans also brought to Europe certain artefacts associated with sun worship. These included new beach and leisurewear fashions – the swimsuit, lounging pyjamas, shorts – uncovering the body – quite the opposite to the pale features encouraged by the Victorians and Edwardians. In environmental terms, this meant 'discovering and exploiting' the beaches of Mallorca, previously largely neglected (Buades, 2004: 43).

> La platja es converteix en el lloc de la festa per excel.lència, on poden despullar-se de la seva intel lectualitat i de la vida industrialitzada, on la roba podrà deixar d'incomodar el cos, on les convencions socials deixen pas a un lleuger llibertinatge. [The beach becomes the place of festivity par excellence, where one can strip off the intellectual and industrial life, leave behind restricting clothes and conventions, taking on a lighter freedom]. (Buades, 2004: 43)

The beach had become a stage or theatre, the symbolic 'other' to urban life in northern Europe, converted from 'the edifying seashore to the pleasure beach', acting as 'an attractor of fantasies and desires, one of those leisure spaces specially designated for the purpose of pleasure and physical gratification for modern city dwellers'. The beach became a constructed space –'a stage, a playground, a site for edification, sexual gratification or simply easy living' (Baerenholdt *et al.*, 2004: 49–51).

This combination of 'sea', 'sand' and 'sunshine' as a tourism resource is primarily, then, a product of the particular values that tourists' cultural traits placed upon relatively constant environmental features. The transitions from one historical period were themselves by no means constant. For example, even as late as the 1940s many working-class visitors to Mallorca did not come for this beach experience but like their

Edwardian predecessors came more to see the 'life' of the island, a sort of ethnographical tourism (Barton, 2005: 199). These perceptions of the physical environment were to change drastically with the advent of mass tourism in the 1950s. By then the focus of practically all tourism had switched to the coast and its beaches, forcing local authorities to ensure that the 'beach' met with the perceptions and desires of visitors, leading to a policy of environmental management that ran contrary to the natural processes of beach formation and development. The physical 'beach' itself became a social construct.

Two main types of beach may be identified in Mallorca: the long but narrow kind fringing large open bays, and the beach at the head of a cove set between short parallel headlands. While the former is open to larger wind and wave action, the latter offers sheltered bathing and accommodation for sun worshippers. Geologically, the seas immediately around the island shelve gently to deeper water, except on the north and west coast, offering safe bathing conditions nearly everywhere. Amongst the larger beach formations are those of the south coast east of Palma – the Platje de Palma and S'Arenal, and the largely undeveloped beach at Es Trenc. In the two northern bays of Alcúdia and Pollença, the 5-km long, narrow beach from Punta Larga de s'Estanyol to Port d'Alcúdia supports a whole string of resorts including Can Picafort. The small cove-like beaches are to be found in two locations: to the west of Palma, primarily in Calvià and includes some of the early resorts such as Calamajor and Palmanova, and the more recent havens of youth culture at Magaluf, Santa Ponça and Paguera. The second group is strung out along the east coast – the calas of Mallorca – including Cala Figuera, Cala d'Or, Cala Murada and Porto Cristo (Photos 2.1 and 2.2). The Bay of Son Servera is something of a cross between the two major beach types in that it is a long, narrow beach stretching north from Punt d'en Amer to Port Vell but set in a large cove and includes the conjoined resorts of Cala Bona and Cala Millor.

Rodríguez Perea *et al.* have written that

> Les platges de les illes balears constitueixen el principal actiú del medi ambient en què es basa l'economic turistica de les Illes... Les platges són espais que pertanyen a sistemes naturals molt fràgils i molt dinàmics. [The beaches of the Balearic Islands constitute the main environmental asset as the basis of the tourist economy ... Of the spaces that relate to natural systems the beaches are the most fragile and dynamic]. (2000: 13)

Their study of beach–dune dynamics and how they operate reveals just how much the modern beach is a creation of engineering and not geomorphic or 'natural' processes. While this is not unique to Mallorca (see Malvarez *et al.*, 2004: 203), some local details are worth noting.

Photo 2.1 Pool and beach in close proximity in Cala d'Or

Photo 2.2 Modern hotel in 1930s planned resort in Cala d'Or

Firstly, the natural beach material of most Mallorcan beaches is not the 'sand' that is seen today but instead about 70% or more is composed of bioclastic material, the bony and chalky remains of small sea creatures. Its production and maintenance is intimately linked to the stands of *Posidonia oceanica* (what is called locally *alga* – seaweed) that lie just offshore on the submerged beach (Figure 2.4). As part of its life cycle, it sheds large amounts of dead leaf material that accumulates on the beaches in winter to form low-cliff-like features that protect the beach from the ravages of winter erosion (Photo 2.3). While tidal variation may be small in the Mediterranean and longshore drift restricted, nonetheless wave damage to beaches in winter is significant. This is compounded by often massive run-off from torrents that literally tear across the many Mallorcan beaches formed at their seaward exits, for example, in the many calas on the east coast (Photo 2.4). The *fullas* (wasted leaves) of *P. oceanica* also help protect this type of loss (Photo 2.5).

Secondly, the natural long profile of Mallorcan beaches shows a range of shadow dunes at the head of the beach. These provide a cumulative source of beach material, but more important they act as a barrier behind the beach itself. Destruction or degradation of this dune system rapidly produces an imbalance between deposition and erosion (Rodríguez Perea *et al.*, 2000: 17). This relationship has easily been upset by mass tourism.

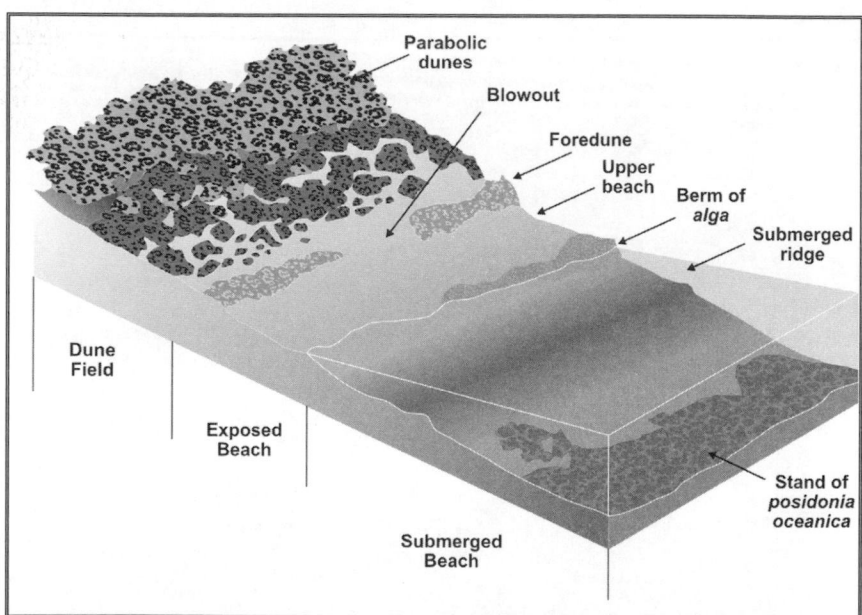

Figure 2.4 Natural process of beach formation
Source: Servera (2004)

Photo 2.3 Regenerated Porto Colom beach damaged by autumn storm

Photo 2.4 Destruction of artificial Calas de Mallorca beach by autumn storms (28 September 2009)

Photo 2.5 Alga giving protection to natural Son Real beach

Many land-based factors have led to this degradation besides the sheer weight of numbers of beach users, including the erection of *chiringuitos/quioscos* (kiosks) on the beach or edge of shadow dunes, promenade construction and other aspects of resort building that affect longshore drift and other natural sea processes, the construction of marinas and the overuse, generally, of the shadow dunes. Amongst the sea-based factors are yacht anchorages in stands of *P. oceanica*, the polluting of sea waters inimical to it and dredging.

Two case studies illustrate some of the problems of beach management. The first is that of Cala Mesquida on the north coast where what was once a wild and exposed beach has been degraded by various agencies over the last 50 years. These led to degeneration of the foredune, giving rise to blowouts, damage to vegetation and increased storm damage from sea and surface waters exacerbated by the loss of *P. oceanica*. Between 1956 and 2005, blowouts increased by 40% and the beach–dune system lost 75% of its vegetation cover (Martin-Prieto *et al.*, 2007). In contrast to the originally difficult-to-access Cala Mesquida, the second example is drawn from the popular beach at Platja de Palma and S'Arenal. Data from 1956 to 1979 show that the beach area grew by 25.54%; between 1979 and 1990, it grew by 15.85% but between 1990 and 2004, it decreased by 14.78%. In the first period, it grew by natural processes. In the second period, the slower rate of growth was largely caused by the construction of a promenade and required beach replenishment. By the third period, replenishment had actually begun to cause erosion (Coll Ramis *et al.*, 2007: 197).

It would appear that early in the history of mass tourism considerable opposition arose to the presence of the rotting remains of *P. oceanica* at the sea's edge and to natural bioclast beach material. The former was perceived as dirty, infested and often smelly and the latter as too rough to lie on. Two solutions were offered: to clean the beaches of *alga* with heavy machinery, often on a daily basis in early and late seasons, and to regenerate the beaches at the beginning of each season with sand dredged out from sea and pumped onto the beach. This is often opposed by environmental pressure groups such as Grup Balear d'Ornitologia i Defensa de la Naturalesa (GOB) who, for example, recently campaigned against the transport of dredged sand from Banyabalfur on the west coast to replenish the beach at Cala Agulla in Capdepera on the north-east coast (Ultima Hora, 16 September 2009).

These two 'solutions' are both temporary: the first has to be under-taken at the beginning of each season often on a massive scale and then at frequent intervals along the summer shoreline. In any case, the scrapers used also compact the sand into 'unnatural' forms. The second is temporary because the artificial sand is simply washed back into the sea by a combination of autumn and spring rainfall often falling in tempestuous storms (e.g. on 28 September 2009 – Photo 2.6) and by

Photo 2.6 Destruction of Calas de Mallorca beach by autumn storms (28 September 2009)

higher tides and winter storm waves. However, both of these actions by municipal authorities – seaweed removal and sand regeneration – will continue to take place. For them there are sound economic reasons – but in defiance of environmental well-being – because the 'beaches' are the important resource for tourists. The benefits far outweigh the costs, which are publicly borne (by the municipalities) in any case but recouped via price rises and taxation paid by locals who, in turn, benefit from tourism's externalities. Where policy action has been more influential has been to restrict building on the coast, especially following the Lei de Costa of 1988, the recent restrictions on marina building and restrictions on *chiringuitos* (beach bars) on the beaches.

Thus, the beach has become what some have called a touristed landscape showing complex relations between the natural landscape and its 'alteration' by being viewed by locals and visitors in different ways (Cartier, 2005: 2). Its

> seduction . . . is about sensory allure of 'sun, sand and sea' and more. It is about the interface of continents and oceans, the potential fear of being faced with the enormity of the water planet, the seduction of being at ends, of earth, land and the impact of urban life. It is about resonances, of sand as the earth's constituent or residual matter, as it

dissolves into the shallows of oceanic vastness. It is about the simultaneity of the sensory: sound of the waves, the sight of the surging water continually resurfacing the beach, the feel of sand on the body, the smell of the marine, the taste of salt in the air. (Cartier, 2005: 14)

Rather exaggerated, perhaps, but it is more about myth than the reality of the physical geography.

It is possible to find unreconstructed beach landscapes such as around Port Vell where the *posidonia* has been left, providing a natural beach behind which can still be seen examples of early tourist summer houses, giving a glimpse into Mallorca's seaside before the masses descended and tourism became industrialised. Similarly along the unspoiled beach at the *finca publica* of Son Real on the north coast, natural beach forms and processes occur (see Photo 2.5).

The mountains

The second physical landscape element, but one that is of much less significance to modern mass tourism, is the mountains that lie to the west along a south-west/north-east axis with an almost parallel line of hills in the far east and the north-east. Only the former – the Tramuntana – might be considered a true mountain landscape with several peaks rising above 1000 m, the highest being Puig Major at 1450 m. The eastern hills – the Levant hills and the hills of Artà – are considerably lower at around 300 m and mostly covered in low scrub and garriga, lacking for the most part the bare rock surfaces of the Tramuntana and with less cover by *Pinus halepensis* (Aleppo pines) and *alzina* (holm oaks).

Historically, in terms of human settlement, the interdigitating valleys and throughways of the Tramuntana proved attractive from the prehistoric period to the early modern period. Although always quite sparsely settled, they provided refuge, were well watered and had rich resources in timber and pastureland, especially attractive for early Arab and Maghrebian immigrants' *alquerias* and *rafals*. The Christian conquerors after 1229 took over these estates, particularly for sheep grazing and the exploitation of the timber resources, especially for charcoal production; the useful timber resources were soon exhausted and in any case the holm oaks and Aleppo pines proved inadequate for most constructional purposes. Even the relatively low population numbers and densities by the beginning of the 20th century were further reduced as urbanisation took hold, especially when the tourism industry began to take off in mid-century.

However, early visitors from northern Europe found the Tramuntana attractive, complementing the heat of the plains and providing the kind of 'Alpine' environment that was beginning to prove inviting elsewhere in mainland Europe, if here only on a miniature scale. Captain (Clayton, 1869: 237) wrote that they were

> a charming blend of savage wilderness and fertile cultivation. In no part of the world can one behold a more complete picture gallery of all the varieties of natural scenery than in the Isle of Majorca ... a mighty ravine, cleft by some great throe of nature, yawned to such an enormous depth that it seemed to disclose the very core of the mountain ... far below arose black towers of rock and broken pinnacles looming in the rising mists. It made the senses reel.

The mountains also had the advantage of being relatively close to Palma – the major tourist base until the 1950s – and so visitors could perhaps hire carriages to take them up into the hills or like J.E. Crawford Flitch journey through the mountains on foot and by mule – 'driving is out of the question', part of a vivid descriptive chapter in the Edwardian period (Crawford Flitch, 1911: chap. 3). In the 1920s, hill walking was already fashionable for the English as well as Spanish visitors; Chamberlin's guide devotes much of its space to this activity (Chamberlin, 1925). By the 1930s, cruise line visitors could take excursions via taxis and charabancs.

In addition to the physical attractiveness of the mountain environments, the Tramuntana was the locus of numerous cultural and historical sites – hilltop castles such as Alaró and monasteries such as that at Lluc.

In the present period, and especially from the 1980s, the mountains of Mallorca have again become an important physical resource for hikers and walkers. There have long been local hill walking societies in Mallorca often associated with church groups, the cult of physical exercise and manliness but the 'discovery' of the mountain trails, especially by German visitors, has led to a 'boom' in this kind of diversified tourism. It is underwritten by guidebooks and maps in order to make this environment more accessible. It, rather like the beaches, has been 'industrialised' by the construction of new pathways or the restoration of old ones, by signposting and interpretation boards. The mountain rescue service provided by the Guardia Civil and the *bomberos* is now on hand with helicopters if the global positioning system systems fail. While the 19th-century visitor may have perceived Tramuntana as somewhat remote and forbidding, the modern hiker sees the mountains as welcoming and, to a large degree, safe. In Chapter 8, this kind of diversified tourism will be discussed further; here, it should be noted that while the physical reality of most of the mountain and hill environments has remained the same, the perception of them by tourists has changed.

The plain: Es Pla

The third landscape element is Es Pla, the central plain of Mallorca made up of 14 municipalities to form a *comarca* (region) of undulating low hills in its central area, cut through by numerous largely dry valleys and at its northern and southern ends fading into the low-lying *marinas*, areas formerly under the sea (Figure 2.3). This is not a region that attracts the mass tourist in any numbers. It is a zone to be glimpsed from the airplane as it descends to land at Son Sant Joan airport or passed through on a coach trip or side tour to the markets of Mallorca's inland towns. In recent years, however, it has attracted more tourist interest primarily as a counterpoint to the excesses of the coastal resorts.

In human landscape terms, it is an area that evolved with the large estates of the original medieval settlers, which, in turn, were subdivided in the 18th and 19th centuries into the *possessions* of the new landed classes and which themselves were further divided into the tiny minifundias of the *pagès* and smallholders.

From the late 1950s, it was an area of high population loss as agriculture declined and its young people moved to the coastal resorts to take up the new employment offered in tourism. However, to the new breed of residential tourist, it offered quiet rurality and low property prices. A landscape of small nucleated towns and isolated farmhouses does not, of course, have the benefit of the sea and suffers from high summer temperatures, hence the proliferation of swimming pools attached to the renovated farms. Many, including many Mallorcans, see this area as the real Mallorca, 'una imatge bucolica i irreal ... comarca de la Mallorca profunda ('an unreal, bucolic image ... the region of fundamental Mallorca') (Picornell & Picornell, 2008: 205). It was also the location for much of the new agrotourism and rural hotel businesses that grew rapidly in the 1990s. Dominated by a wide variety of sedimentary landforms, it produces a rich variety of soil types and colours ranging from the dark reds of the terra rossa to the almost white soils of the chalks, one of its visual attractions. The abandonment of farming over much of the central area has seen a remarkable rise in tree cover, especially Aleppo pines, and scrub. The areas of long neglected almond and carobs present another aspect of this air of decline. Mallorcans bemoan the loss of this heartland, but it is a landscape of almost romantic proportions attractive to the German or British second home owner offering them a tranquil tourist resource that contrasts strongly with the crowded coastline (Ballester *et al.*, 1993: 117–118). It comes at a price, of course. Increases in land values have excluded many locals although around Palma there is now considerable evidence of rururbanisation under demographic pressure and a perceived need to escape big city living. All the coastal settlements, and especially the area of Greater Palma, now exploit this zone's resources themselves. Water, in particular, is now drawn from the aquifers beneath Es Pla, for example, from Sa Costera, S'Estremera and the deep wells at Sa Marineta – all this in addition to the supplies drawn by the more affluent locals and second home owners for their many baths, swimming pools, gardens and golf courses. What agriculture remains is increasingly focused in this area, and irrigation remains by far the largest consumer of water.

In many ways, tourism has created two Mallorcas – the coastal coral reef and within it the lagoon of the changing rural interior. The perception of this area as a resource has changed. Once seen as backward and likely to be abandoned, planners in the mid-1990s suggested limiting house building and allowing agriculture to continue to decline, but both of these activities offer perhaps the only solutions to Es Pla's economic and social problems. It was perceived as a kind of Arcadia or paradise, hardly likely to improve the quality of life of those who lived and earned a living there. Regrettably, under Parti Popular the powerful road lobby saw this area as one to be crossed by new high-speed roads (*autovia*). The geographer Climent Picornell sees it as a region in crisis in terms of traditional activities and sees the danger of using tourism to bring new life to the area if it becomes an ethnological park – a 'reserve' along the disastrous lines of the Native North Americans' experience; worse still would be the creation of the area as a theme park (Picornell & Picornell, 2008: 23). What is required is a means of developing an interaction between tourists and the landscape and heritage of Es Pla that benefits both parties as well as supporting the physical and human resource base for new forms of agriculture. To Picornell the idea of 'Mallorca profunda' is a dangerous fiction that could undermine the proper survival of this area (Picornell & Picornell, 2008: 37).

Albufera

A final physical region that has possibilities for tourist use but that was considered in the 19th century to have agricultural potential is the Albufera, Mallorca's only extensive area of wetland. Composed primarily of standing water and reeds it is the remnant of a lagoon-like feature that is cut off from the sea by dune formations. Such wetlands are curiously characteristic of the Mediterranean coastlands. In Mallorca's case, the Albufera is located in the Bay of Alcúdia with a smaller version – the Albufereta – in its neighbour, Pollença Bay. Originally a productive source of wildfowl and fish, attempts were made in the late 19th century to drain large parts of it for, amongst other crops, rice. The English company that cut canals and introduced steam-powered drainage had little success. In the 1860s, 30%–40% of the Albufera's surface was open water; today, it has been reduced to about 3%. By the

1950s, it had begun to revert to its more natural state. It was not until the rise of the environmental movement in the middle part of the last century that the Albufera's ecological importance and its potential for controlled tourist access began to change once again the value that society put on its landscape. Now a natural park, it is well laid out with footpaths, bird watching hides and refuges for wildlife that are proving attractive to the more active tourist. Unfortunately, the hotel building industry of the 1970s did not ascribe to this area the same ecological value, and the local authorities (Muro and Sa Pobla) permitted a great deal of resort development on the protective sand dune coastal strip, eroding an important element of the whole wetland system (Martínez & Mayol, 1995). This is a final example of not only how the physical geography and the landscape have been altered by human action, but more important it is a good illustration of changing perceptual values – and uses – in an age now dominated by tourism.

Chapter 3

The Historical Development of the Tourism Industry from the late 19th Century to the mid-1950s

Introduction: A Complex of Historical Factors

The origins of the modern tourism industry have to be searched for from amongst a myriad of factors related to geography and history, to sociology, to politics and economics and to an understanding of business development both individual and corporate. Some of these may be place specific, others of more universal application. In the case of Mallorca, the context for the period up to the mid-1950s was characterised by the changing nature of European society under the twin influences of urbanisation and industrialisation combined with the rapidly increasing population. As the demographics of Europe changed from the depression of the 1870s to the post–Second World War 'boom' in most countries, so did the nature of work, releasing more leisure and non-work time. We will see that this release had a quite marked class basis. The influence of early upper middle-class travellers may have been important in collecting information about the distant parts of Europe such as the Balearic Islands nearer the beginning of this long period, but the 'masses' required to sustain a long-distance tourism industry did not materialise until the 1950s and 1960s. It will be important to see whether there really is any systematic link between these early travellers and the later package tour holidaymakers. However, the growing certainty of a ready supply of potential tourists was by no means a guarantee that their demands would or could be met. For Mallorca there needed to be not only a resource that potential tourists might wish to consume, whether that be environmental or cultural, but also a means of mediating between supply and demand had to be developed, including a transportation system, an accommodation system and an infrastructure that could meet other demands or develop new 'attractions'. Small groups or individual visitors made quite different demands from the 'masses' that were to materialise after the Second World War. Mass tourism like most other tertiary activities was a business, a product of mid-to-late capitalism, that had to be developed in both the geographical centre of tourists' origins and in their destinations. By definition, it was an economic activity that responded positively to economies of scale, but above all it was the speed with which Mallorca's business community and public servants reacted to the possibilities raised by European travel agents and tour operators.

To imagine that because a limited number of well-off visitors went in search of their respective 'paradises' from the Age of Enlightenment onwards and brought back reports of untold cultural riches that would, in turn, attract other visitors is to oversimplify grossly a complex set of variables that themselves changed over time (Fiol, 1992). This study of Mallorca, one such 'place' amongst many such 'places', will try to show how these variables interacted to produce one of the richest and densest tourist environments in the world, something of a laboratory in which various tourism experiments were tried, which many other 'places' copied, learned valuable lessons from and went on to develop their own unique industries. In addition, those who managed the industry learned similar lessons whether they were based in Britain, Germany and Scandinavia, peninsular Spain or Mallorca itself. Indeed, in a later chapter we will seek to show that the Balearic model of tourism development has, in a more globalised world, been widely adopted and often led by Mallorcan companies (see Chapter 8).

It will be convenient to divide this long period of tourist development into a number of historic phases that illustrate the variety of socio-economic and geographical forces outlined above. None of these phases is 'watertight'; they all overlap and intertwine in different places and at different times especially when examined as spatial processes despite, or perhaps because of, the inexorable rise in tourist numbers visiting Mallorca, from a few thousand in the mid-to-late 19th century to more than 9 million today. The island of Mallorca has not grown in size in these 150 years and so perhaps the most remarkable statistics of change are those of tourist density.

At a simple level were a series of 'pull' factors, indigenous characteristics that were to attract tourists, against which must be posited the exogenous or 'push' factors that made the outflow of visitors possible. While there may be debate about the term 'mass tourism' (Walton, 2009: 785), it was not invented in Mallorca but rather it grew and was refined primarily in Great Britain, the product of a small number of British entrepreneurs and the companies and firms that they established and transplanted to the Mallorcan environment (both physical and economic) where it flourished in partnership with local businesses in the post–Second World War era.

Lastly, in this early period before about 1900 we will see the beginning of the processes of transformation operating in Mallorca – of its landscape and environment, its social structures and values and its economy. Tourism may be described as an agent of systemic change, leading eventually towards a post- and later non-touristic economy. In a later chapter, we will show that the classic Butler model of 1980 is only partially useful in describing and explaining these changes in Mallorca. It is important therefore to begin our analysis of the development of the tourism industry with some understanding of what preceded it in the Mallorcan economy. This, it will be proposed, alongside the description and analysis of the island environment already undertaken, was part of the complex set of factors that turned the island towards tourism. An understanding of the island economy at the end of the 19th century provides the necessary context for the development of tourism in the 20th century as an all-embracing tertiary activity.

Mallorca in the late 19th Century: A Changing Economy

The development of a growing, organised tourism industry in Mallorca, as opposed to the island simply being a destination for travellers or seasonal visitors, might be seen as a response to a relatively poor economy based on agriculture, industry and trade. While it is true to say that these latter activities expanded considerably in 19th-century Mallorca, they remained at the mercy of fluctuating world markets and prices, often in competition with other Mediterranean producers. Mallorca had always had great difficulty feeding itself and had to import substantial amount of food for centuries, especially bread grains, requiring considerable foreign exchange. Export earnings are a continuing theme in this early history. Isolation and transport costs, so often cited as being responsible for Mallorca's so-called backwardness, were not really a major hindrance to the growth of the island economy; they were features of all the Mediterranean islands and were more a product of the poor and unreliable environment and the archaic organisation of land-holding and farming rather than of an unfortunate geographical location (Manera, 2001: 17). The island's population grew under the stimulation of these economic advances, but the recessions of the late 19th century resulted in a considerable out-migration of population to Latin America and elsewhere. Traditional underemployment in the countryside later became paralleled by unemployment in Palma and other towns.

Looked at in more detail, Mallorca's economy by the mid-1850s was at last beginning to break away from its feudal roots. The trading role that the island had specialised in was increasingly complemented by improvements in agriculture and industry. Land reform beginning in the 1830s (*desamorticazion* – disentailment) saw more entrepreneurial forms of farming develop under the influence of a new land-owning class coming out from the city and acquiring estates – *possessions* – from the original feudal aristocracy. This latter group had controlled farming in a most negligent manner for centuries, but from the 1840s large areas of land were cleared and enclosed by the rectilinear stonewalls that form such an important element in the contemporary landscape. This creation of fields had gone on since Arabic and Berber times around the alqueries and rafals, many of which were to form the pre-urban nuclei of the island's planned towns of the Ordinances of 1300. This progressed slowly through medieval and early modern times but accelerated from the mid-18th century onwards. Much of this 'enclosure movement' was undertaken for the new landlord class by *roters*, the poorest of all agricultural workers. In addition, a new class of small farmers (*pagès*) grew out of the former landless class to occupy a large number of very small and sometimes scattered holdings but which could be fairly intensively farmed on a share-cropping basis. Wheat was always the staple crop although many of the new landowners and small farmers did grow pulses and beans as an alternative source of carbohydrates (Manera, 2001: 113).

In the 18th century, the cash crop that gave a late stimulus to the aristocratic landowners had been olives, especially in the mountainous area to the north and west of the island. Besides being used for cooking, lighting and food products, olive oil found a major outlet in the soap boiling industry of Marseilles. Viticulture had always played some role in the Mallorcan agricultural economy, but its expansion on more commercial lines was instigated by a rising demand for stronger drinks such as brandy, and industrial alcohol. However, the period of most rapid growth – indeed, it might be described as a veritable 'boom' – occurred after the 1870s when phylloxera struck down production in the south of France. Mallorcan growers rapidly responded to French misfortunes by expanding production, shipping out barrels from newly developed ports such as Porto Colom on the east coast. However, by the 1890s, phylloxera had reached Mallorca and local viticulture was more than decimated, as it had been further north. France meanwhile had begun to recover as disease-resistant vines were developed on American stocks and so French production of red wines was rising again, compounding Mallorca's woes (Manera, 2006: 67–73, 84–100).

Mallorcan soils were, and are, relatively poor and in the days of horse and mule power difficult to plough, which meant that traditional arable farming in much of the countryside was characterised by low yields, especially when coupled with the notoriously unreliable rainfall in the growing season and so the most revolutionary changes to the countryside occurred with the widespread commercial adoption of a trilogy of tree crops: figs, almonds and carobs. The growing urban industrial economies of northern and Western Europe provided substantial demand for the fruits of these crops for industrial processing, sources of raw material for oil, food products and animal feed (Manera, 2006: 74–80).

The landscape changes brought about by agricultural change were to have a significant effect on later tourism in two ways: they constructed a more domesticated and better maintained countryside that later visitors found accessible and attractive – a distraction from the sea, sun and sand elements – and in the almond tree produce, with its sea of pink blossom, a landscape element that each February became part of the 'pretty picture' beloved of a generation of the rather more romantic kind of early visitor.

Mallorcan industrial development in the 19th and early 20th centuries was not to provide the same kind of residual features for later generations of tourists. However, certain products of industry were to become important for tourist retailing, such as leather goods, glass, textiles and pottery, although today the majority of these items are often made elsewhere in Spain, in north Africa and in this era of globalisation, in many Third World countries, often masquerading under a 'local' label. Some of the factories and workshops that produced these goods have acted as venues for side tours, again as distractions from the beach.

At this point it is important to try to assess the role of industrial development in Mallorca in pre–mass tourism days as part of the economic development of the island that, because of its relatively low level of development in terms of supporting the population with jobs, goods or sufficient overseas earnings, helped, in part, to turn the economy towards tourism.

The textile industry dates from the middle ages and from then until the 19th century was characterised by small-scale, local production of woollens, linens and cottons. Much of this production took place in towns rather than the countryside but was nonetheless a 'cottage' industry. Originally water powered, with the coming of steam power from the 1840s, a factory system soon replaced some of the small workshops with increased mechanisation and a consequent reduction in demand for labour. A bi-polar structure of a large number of small workshops existed alongside a small number of larger factories and mills. The same processes could be seen at work in metal work and the woodwork industries. The footwear industry also has a long history in Mallorca, but output was concentrated in urban areas, especially Palma, again primarily in small workshops; mechanisation arrived relatively later (Escartin, 2001; Roca & Umbert, 1990: 106). Tables 3.1 and 3.2 give an idea of employment structures in the late 19th century.

Table 3.1 Evolution of active population by sector in Balears (%)

	Primary	Secondary	Tertiary
1787	53	15	32
1860	68	13	19
1900	70	15	15

Table 3.2 Distribution of industry by type in the 19th century (to nearest%)

Industry	Year		
	1799	1856*	1900*
Food	—	53	35
Textiles	43	16	22
Distilling	23	—	—
Metal work	23	< 1	6
Leather, shoes etc.	3	8	4
Pottery, glass	3	7	5
Woodwork	2	< 1	5
Chemicals	—	14	8
Paper	—	< 1	2
Construction	—	—	—
Total	100	100	100

*Figures for all Balears; rest Mallorca.
Source: Adapted from Manera and Petrus Bey (1991)

At the same time as the structure of the economy was shifting from the primary to the secondary sector and as mechanisation was reducing the demand for labour, two other variables were beginning to affect the economy: emigration and the loss of what remained of the Spanish Empire. In the last decades of the 19th century and into the next, a combination of rising demographic pressure and falling employment opportunities forced Mallorcans to seek new lives abroad, especially in the Americas. The population of Mallorca rose from 204,000 in 1857 to 250,000 in 1887 to 270,000 by 1920. Pere Salvà has calculated that between 1878 and 1910 Mallorca had a negative population balance of 41,453 (Salvà, 1992: 406–407). By the end of the 19th century, agriculture was no longer able to support the rural population and numbers did not increase in the last two decades of the century. Despite increased urbanisation and industrialisation – the usual solution to rural depopulation – and Palma contained about a quarter of the island's people – new economic activities did not spread widely enough to absorb these numbers. The only alternative was emigration to the Spanish-speaking parts of Latin America. Mallorca had often lost population abroad in the past, mostly to the Peninsula, France and the Maghreb, but the loss from the 1890s to the 1920s was catastrophic, especially after the rising economic optimism of the 1850s to 1880s. The poorest rural workers were hard hit, but they often found it difficult to fund emigration; a surprising number of the emigrants were from Palma, the major city.

This then was the economic background of the late 19th century against which we have to view the rise of the tourism industry: increasing industrialisation and urbanisation, variable export earnings, increasing imports, reducing job opportunities especially for an unskilled and largely illiterate workforce and rising emigration. Added to this was a considerable disparity in average incomes, with a small but immensely rich – and often absent – aristocracy at the top, a growing sector of professional classes particularly in banking, medicine and the law and a large, poorly paid class made up of small farmers, landless farm labourers and a small but increasing urban proletariat at the bottom. The urban structure was highly primate, dominated by Palma, which in 1900 had a population of 64,000, about 25% of the island's total. However, it was the events of the last two decades of the century that were to throw into sharp relief the precarious nature of the Mallorcan economy, notably the collapse of the local wine trade referred to above, and in many ways more important, and on a larger scale, the longer-term effect of the loss of the remnants of the Spanish Empire in 1898 in the Caribbean and the East Indies. We have made reference to population loss through outward migration at this time and clearly if the island's people were to survive and prosper and not face the even greater losses that were a feature of so many Mediterranean islands at this time, then a different economic approach was needed.

Although the later parts of the 1800s witnessed these economic upheavals, a positive survival was a well-connected entrepreneurial class of businessmen. In a male-dominated society, they had learned from their various experiences of industrial and commercial development in manufacturing and trade. Many were quite well travelled in the Peninsula and in Europe more widely. They had an enquiring outlook, had observed changes in Europe's economic activity and social structures and were well versed in the role of risk in business. Mallorca may not have been a particularly inventive society, but it was well used to adopting and adapting new technologies and innovations into the ways of conducting business.

The Beginnings of Commercial Tourism

Throughout the 19th century, the number and range of visitors to the island had been increasing. Nearly all were at least middle class, and many belonged to the artistic community. Barceló and Frontera give an impressive list that includes naval and military visitors, British, French and German travellers and sojourners, many of whom recorded their impression of landscape, people and culture; some British examples were noted in Chapter 1. Through them knowledge about Mallorca was spread amongst their like-minded contemporaries whether they were artists such as Sorolla and Rusiñol, writers such as Gaston Vuillers and Charles Wood and scientists such as the geologist Martel and the German zoologist Pagenstechen for whom the influential Archduke Ludwig Salvator, referred to in Chapter 1, often acted as host (Barceló & Frontera, 2000: 16–20). Buades describes the tourists of this pre–First World War 'belle epoque' era as 'aristocratic', a social group who demanded a relatively rapid access to European attractions and above all comfortable service. Such tourist movements became linked, then, to fast steam ship services from the United States and rapid rail transit across Europe, accompanied in Mallorca's case by an improving ferry service from the Peninsula or the South of France. Certain technical advances in banking also aided foreign travel at this time: the traveller's cheque and American Express money orders. Modernisation of aspects of the island occurred in these early years, including the introduction of electricity and the spread of motor vehicles, accompanied by new roads and rail lines such as the new line to Sóller and the extension of the Manacor line to Artà. In the days before beach holidays, these upper middle-class tourists were more interested in the mountains and the island's natural phenomena but usually based themselves in Palma and so these new routeways were designed to give better access to sites such as Banyabulfar's terraces, Artà's caves to the north of Canyamel and eventually to the wild cliffed coasts of the north and west and Mallorca's two alternative northern bays, Pollença and Alcúdia. While much of Western Europe was embroiled in the massacres of the First World War, Spain remained neutral and its Mediterranean islands acted as a haven of peace, to many a veritable paradise, characteristics various international expositions were not slow to publicise.

It is important to remember that while Mallorca and the Balearic Islands were geographically peripheral to the Spanish Peninsula, they were not isolated from Spain's political movements, especially those relating to tourism. Spain was seen by the north European organisers of this growing activity as being more 'oriental' than 'European' with an exotic appeal (Walton, 2002: 123). There was a move for Spain to try to find its place in the modern world. Tourism was seen by some as one vehicle for achieving this, but the country was perceived as lacking the services and facilities that the modern traveller required, especially decent hotels. If tourism was to flourish, efforts had to be made to improve this image. To this end, a National Commission was set up in 1905 to promote 'artistic and recreational excursions for the foreign public', emphasising the country's climatic advantages as well as local culture and the rich heritage. This was followed by a Royal Tourism

Commissariat in 1911 that sponsored the 'Sunny Spain' exhibition in London in 1914 (Gonzáles Morales, 2005:18). In some ways, Mallorcan entrepreneurs had pre-empted some of this development with their identification of 'La industria de los foresteros' in 1903, indicative of the reciprocity between island and mainland.

Within this emerging national context, many local observers had witnessed the growing numbers of outsiders visiting the island and settling there during the last third of the 19th century and Mallorcan writers and intellectuals began to consider them as a new source of economic activity. The first of an influential trio of commentators was Miguel dels Sants Oliver Tolra who wrote a series of articles in a local newspaper, *La Almudaina*, in 1890 that was later to emerge as a book, *From the Terrace* (1891). This exposed the reigning provincialism and the economic malaise of the Balearic Islands but as an antidote suggested encouraging tourism as an economic activity based on the island's natural resources and climate. He believed that the existing type of middle-class traveller was looking for 'something new, something unaffected, something not yet explored'. Noting that all too soon modern business would transform the island via tourism – building roads, opening hotels, exploiting the natural caves etc. – he advised that Mallorca should take advantage of this business opportunity before the very business itself altered the very thing that the visitors would come to see, a very prescient view!

To add to their enjoyment and understanding of Mallorca, a number of early travel guides had been written by locals such as Miguel des Sants Oliver in the 1890s. In 1891, Pere d' Alcàntara Pena Nicolau, a writer and engineer, published a *Handy Guide to the Balearics (Guia manual de las Islas Baleares)*, noting that the new breed of visitors would need a good guidebook that should also be designed to publicise the islands' attractions. In it he drew attention to the sort of features that might attract visitors –

> Sus paisatges, seus puntos de visitar son en gran manera pintorescos. Sus terrenos se cultivan con aquel cuidado y esmero que notamos en las jardins particulares. Los palacios del antigua nobleza continien gran numero de objectos aristicos... y la historis... de las islas esta llena de interes, especialmente para los hijos de le gran Bretana. [Its landscapes, its noteworthy places are very picturesque. The land is cultivated with the same care and attention that we observe in private gardens. The palaces of the old aristocracy contain many art objects... and the history of the islands is full of interest, especially for the sons of Great Britain.]

Like others he observed that the excessive government bureaucracy associated with travel for foreign visitors needed reducing.

A third key work, *La indústria de la foresteros*, often translated as *The Industry of Foreigners*, was by Bartomeu Amengual Andreu (1866–1961). He believed that catering to overseas travellers could be a profitable and respectable industry. Foreign visitors, he observed, make greater demands than do Spanish ones: 'not for them the camp bed and the cabbage soup (*sopas de Mallorca*)'. Unless better facilities were provided, early visitors would soon discourage their fellow countrymen from following in their footsteps. He suggested that tourism was not just a business for hotel and cafe owners; the state – at all levels – had to be involved in order to take full economic advantage. In effect, he noted what economists would call the multiplier effect and so tourism would eventually affect many sectors. One of his principal recommendations was the creation of an organisation to promote and control tourism in Mallorca. This was to emerge as the Foment del Turisme founded in 1905, the first such organisation in Spain. Run by business people and cultured professionals, it was set up initially to improve conditions in Palma, to publicise the island as a visitor destination, to improve accommodation and island travel, to agree tariffs and prices for goods and services and to provide itineraries and guides. In fact, its main objective was promoting sightseeing – providing the gaze – so important to early middle-class visitors. Its aims were revised in 1927 (Serra & Company, 2000: 70–86).

In the first 30 years of the new century, three significant developments were to translate these economic and marketing ideas into something approaching a modern tourism industry as these intellectuals had hoped. One was the luxury hotel, a second was the development of totally new 'resorts' and thirdly the growth of, for Mallorca, a new form of capitalism based on the power of a banking and financial oligarchy that was very active in acquiring land. Indeed, the latter was probably the means whereby the first two developments could be achieved. It is worth dealing with it first.

As late as the 1930s, agriculture remained the mainstay of the economy but manufacturing industry was growing in the early part of the new century, including for the domestic market, with services proliferating thanks to tourism. But farming now required less labour, leading to a rural to urban movement and so by 1930 Palma contained about 30% of Mallorca's population. Cities are always the locus for a tertiary economy.

Two key figures at this time were Juan March and the Alzamora family. March developed the energy industries of the island, including its first oil refinery and petrol company (CAMPSA); he was also into shipbuilding and with American help (ITT) set up Spain's national telephone company, Telefónica. In 1926, he founded the Banca March, which was to become the source of much of his future wealth and political influence (Roca, 2006).

The Alzamora family originally made its money from the sale and processing of agricultural products. Its family members held a series of important positions in the tourism industry through most of the 20th century, including providing the first president of Fomento. Like March and his followers, they too supported right-of-centre political parties in the period before the Spanish Civil War. They also became important in banking circles.

The 1920 and 1930s saw another large-scale break-up of the large aristocratic estates principally caused by a series of investment and market failures that led to selling land with the help of the bankers. At the same time, the ancient notions of inheritance had broken down and so passing on huge estates to heirs became impossible; they were forced to sell. In Calvià, for example, three old families controlled about one third of the land – much of it fronted onto the sea, including the Marques de Romana's estates at Sa Ponça. To the east of Palma, the Platja de Palma, a coastal zone approximately 200 m deep and 2 km long, was acquired by these banking interests, the site of one of the most intensive hotel developments in Spain later. To the west (Calvià), some estates had been sold off earlier; for example, Paguera was bought in 1886 by Waring, the British engineer who had started to drain Albufera. The death of the Marques of Torre led his heirs to sell off other lands although many were sold in small parcels to small farmers; many of these were eventually bought out through a process of 'acquisition' and so by the mid-1930s much of Calvià's coastline was under the control of bankers (Buades, 2004: 50).

By the 1920s, important and high-quality hotels were added to the resource base, primarily in and around Palma. Domenech's Art Nouveau Gran Hotel of 1903 was the first followed by a series of de luxe hotels built in Palma, Calvià and Andratx municipalities. Most were built on land acquired relatively cheaply from the hard-up nobility whose landed estates tended to turn their backs on the coastal margins. For example, the Mediterraneo was built on the finca Barra d'Or in1923 and the Hotel Victoria on the finca Sa Sabater in 1928. Further along the south coast, the Principe Alfonso Hotel was founded in Cala Major, the Hotel Playas (1928) and the Malgrat (1932), both in Paguera. (Seguí Aznar, 2001: 41).

Some developers sought more remote and often romantic locations rather than the 'suburban' concentrations near the island's capital. These included the Hotel Forementor built in 1928. It was developed by the Argentinian Adan Diehl (1891–1952). He, with others, acquired land in the Formentor peninsula overlooking the Bay of Pollença, an area not properly accessible by road until the 1950s and so clients were often brought in by boat or seaplane. The romanticism of this area perhaps appealed to Diehl via his connections with Argentine poets and artists. However, whatever his inclinations the hotel had to become a strictly commercial venture and soon it was acquired by his banking creditors. These high-class hotels were an important early resource for tourism, but they were to a large extent class defined. Gradually, the number of hotels increased and so by the mid-1930s there were more than 130 on the island.

To many less affluent visitors, Mallorca was perceived as being a land of friendly 'natives', '... in an old European setting with none of the heritage of imperialism, slavery, racial stereotyping and economic, environmental and sexual exploitation that complicate the pursuit of paradise ...' (Walton, 2002: 183). The more luxurious hotels were certainly not aimed at the artistic colony that had come up at Port de Pollença, which the English writer Douglas Goldring had observed as one of the 'pioneers of "Tourism"'. He saw it in 1922 as an unspoiled village whose inhabitants – technically 'Anarchist-Communists' – ran their fishing industry on cooperative lines. Amongst the 'bohemians' were an elderly Spanish painter, a Danish baron who wrote poetry and translated Oscar Wilde and two Danish artists (Goldring, 1946: 187–188). However, a wider artistic community was also beginning to discover the attractions of Mallorca. They looked for relatively cheap hotels such as the three fondas Goldring found in Port de Pollença or accommodation in similarly quiet locations, usually on long seasonal rentals, often seeking to benefit from the winter climate. While it would be an exaggeration to describe them as artistic 'colonies' as were beginning to arise in, for example, the south of France or Tuscany, the low cost of living, the good light and the tolerance of a bohemian lifestyle by the Mallorcans all added to their attraction for painters and poets. This social group did little for the island's economy but a great deal for its publicity, an element that was to contribute to the eventual development of larger-scale – but pre-mass – tourism.

To this group might be added the ex-colonials (of Spanish, British and French Empires that were declining by the 1930s) where a colonial administrator and his wife could find cheap property and cheap servants, enabling them to maintain a lifestyle not too dissimilar from that of the Malay rubber plantation or the Cuban sugar fields. John Walton has described one such 'colony' – El Terreno to the west of Palma – as 'a mixed community of spinster annuitants, frisky divorcees, remittance men and hard-drinking army officers...', initially a 'paradise' but one soon corrupted by the 500 or so British and American sojourners overwintering there in the 1920s (Walton, 2002: 121). Douglas Goldring, writing after the Second World War, remembered why El Terreno in the 1920s before the American colonisation reminded him why 'the more cretinous section of our Conservative middle class supported General

Franco' (Goldring, 1946: 186). By the 1930s, *Time magazine*'s correspondent could write:

> Years ago Mallorca was discovered by a few Britishers who like to dress for dinner in semitropical climates. They encouraged Mallorcans to keep prices amazingly low ($1 a day for hotel room & meals). They swam staidly in the little blue bays, played tennis at the Royal Lawn Tennis Club, in El Terreno, swank suburb of medieval Palma. But in 1931 the peseta sank to a new low and a new horde overran Mallorca: U.S. hard-drinkers who wanted to live like characters in a novel by Ernest Hemingway. They set up their own bars in Mallorca's famed caves. They started a fad of imitating a peacock's screech, slept all day, screeched like peacocks all night. Tourist prices began to skid upward. Travel publicity brought new thousands of law-abiding U.S. tourists, many of whom stayed to open their own shops, restaurants, travel bureaus and pensions. (*Time Magazine*, 24 July 1933)

It was the Americans who brought to the island many of the features of leisure culture that was already sweeping their own country, especially sun worship and the acquisition of a tan as opposed to the pallor favoured by an earlier generation (Buades, 2004: 42). Coco Chanel introduced daring swimwear much to the chagrin of the very conservative Catholic hierarchy (Pack, 2006(b): 661). The main point here, however, is the effect of this cultural shift on the geography of tourism. Whereas the Edwardians had sought out the mountains and the historic or folkloric sights, a new more hedonistic generation focused on *The Beach*. The long stretches of sandy coastline and coves and calas now became the focus of attention, setting in train the beginnings of what was to become later a massive impetus of urban development located on the coasts to the east and west of Palma, on the east coast and in the northern bays (Photo 3.1).

At the same time as the population of Palma grew, pressure on the old city built up and the rising bourgeoisie began to look outwards to suburban living as was happening in so many north European cities in the 1920s and 1930s. Many of the new developments began to perform a dual function, part urban overspill from Palma and part tourist resort. New purpose-built tourist settlements on green field sites, especially in the east and north of island, also began to come up at this time. Seguí has usefully chosen to describe all these new forms of urban development under the heading *ocio* or leisure (Seguí Aznar, 2001). Examining some of them in a little more detail will shed light on some of the processes involved and the patterns produced.

El Terreno, to the east of Palma around the bay, is a good example of the suburb–resort combination. It began life as a zone of summer houses for the middle classes and small proprietors of Palma. It developed in the

Photo 3.1 Relict feature from 1930s in Platje de Palma

1880s on land sold at ridiculously low prices as a series of small houses (*casetes*) painted white, yellow and blue spreading out from Palma's fringe, especially after the demolition of the city's walls in 1901. El Terreno began to be transformed, thanks to the influence of foreigners such as the English in the winter season, renting houses, opening a Protestant church and generally developing an ex-patriot community (Walton, 2005). Added to the area were luxury hotels such as the Hotel Mediterráneo in 1923 and the Victoria Regina in 1928. From about 1925, the area began to be converted into a residential zone particularly between 1930 and 1936 with the construction of blocks of apartments, losing its original identity as it became absorbed into a wider Palma. Barceló identifies a number of housing styles beginning with detached houses in gardens, set in parallel streets, built between about 1880 and the turn of the century. In the second phase from the 1920s came small houses and apartments eventually turning the area into a suburb of Palma but with a fringe of hotels on its seaward side, its nature irrevocably changed with the opening of the Paseo Maritimo at the end of the 1950s (Barceló, 1963).

Another not dissimilar development was La Cuidad Jardin de Palma, a private venture that began in 1918 in Coll d'en Rabassa and included sea bathing facilities, a hotel, summer homes and suburban housing.

Other urbanisations designed on similar principles also date from the 1920s and 1930s, although increasingly many were to be located at a distance from Palma, thus aiding the geographical spread of tourism's infrastructure on the coast. These included San Antonio de la Playa at Can Pastilla, Palmanova, and Portals Nous in the south and Playa de Alcúdia in the north. They were usually planned on green field sites often on former farmland on the less-than-profitable agricultural estates. Their planning took account of the local conditions of relief, climate and vegetation. Many were designed for second homes for the middle classes to be occupied only in the summer. Usually, they had an axial road or a square orientated towards the beach. Housing, often Ibizan in style, consisted of single-family dwellings, but there would also be a hotel and sports and recreation facilities. Amongst examples of this movement are Cala d'Or, Santa Ponça and Palmanova.

For many of the more suburban locations, the seaside was an almost incidental characteristic but by the 1920s beach culture was beginning to turn away from some of the medical and social prohibitions that were imposed in the 19th century, when sun and sand were said to be bad for the body, to sunbathing, even nudity, and hours spent on the beach. The beach began to take on a new socially constructed meaning at about this time. It changed from the utilitarian seashore of the late 1800s to the pleasure beach; it became for many of the visitors to Mallorca in the 1920s and 1930s something of a theatre, as we showed in Chapter 2. Early on the beach became the geographical starting point for many of these new urban developments. The Mediterranean 'season' was no longer the cooler period from October to April but high summer (Lenček & Bosker, 1999: 198–204). In the United States, seaside settlement planning was very much the product of the Art Deco movement but in Mallorca it was more clearly influenced by the garden suburb notions of Unwin and Parker prevalent in Britain at a slightly earlier period. It can also be seen as containing the seeds of Ebenezer Howard's planning ideas with its development separate from Palma's nucleus (Seguí Aznar, 2001: 38–40). At their heart, however, was the politico-economic process of land acquisition that Buades drew our attention to.

Seguí's research shows that Cala D'Or, in Santanyí municipality, for example was developed by an Ibizan antique dealer Josep Costa Ferrer (1876–1971) who acquired land in 1933 that included three *cales*, small secluded and safe beaches between headlands; these were to prove key to its later expansion (Figure 3.1). Initially, the urbanisation was planned to have two 'squares' – a plaza-like open space and a park. Amongst the first purchasers of properties were theatre designers, film directors, the architect Bellini, the wife of Rudolf Valentino and many artists. In 1934, the Belgian Van Crainest obtained permission to build a small hotel of 30 bedrooms near the Calo de Ses Dones. His artist compatriot, the

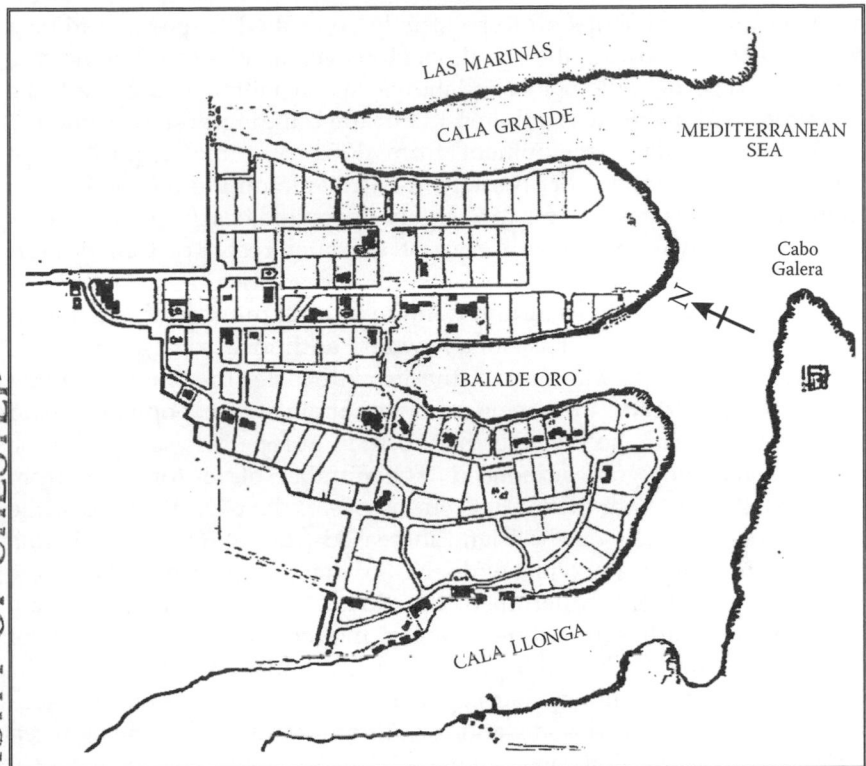

Figure 3.1 Map of Cala d'Or
Source: Seguí i Aznar

painter Vanbergh, acquired land adjoining Cala Llonga where he built the first marina. Later Ses Marines urbanisation was built on land between Cala Gran and Cala Esmeralda although development ceased with the outbreak of the Civil War. Its plan differed from the two previous urbanisations by having a much less formal layout having irregular building plots with a large park of pine trees in the north next to Cala Esmeralda. Eventually, these settlements coalesced and expanded into the present day Cala d'Or, surely a made-up name to attract tourists.

Another example is Palmanova, began in 1934 by promoter Llorenç Roses and architect Josep Goday i Cassias. Part of the original finca de Ses Planes between the Andratx to Palma road and the sea, it was to be given over to a model garden town, again relating the architecture and layout to the countryside and the sea. The main axis would be from the natural promontory where a large hotel would be located, to a central plaza that could be used for outdoor events. Features would include houses with gardens, sports facilities, public buildings, a hotel, a church, a school and local shops all to be strictly controlled. There was to be a *paseo maritima* following the coastline. However, as was so often the case with this type of settlement in Mallorca, when Palmanova came to be built, the road layout was adhered to but the building designs were not.

One spectacular development from 1932 was the acquisition of 3 million square metres of an estate at Son Bauló on the Bay of Alcúdia in the north. This gave access to a largely unspoiled beach 10 km long. Its development was to form the core of the later resort of Can Picafort (Buades, 2004: 67).

Could this relatively small and definitely select number of settlements be expanded to form the basis of a new and additional kind of economic activity? In the interwar period, the growth in leisure time including holidays was already increasing amongst Spain's own population and the boom in seaside holidays further north in Europe associated with the railways had not gone unnoticed. The number of visitors rose from 36,000 in 1930 to more than 90,000 on the eve of the Civil War and while the number of hotels on the island increased from 88 to 131 the Palma municipality still retained just under half of them. Importantly, these new hotels such as the Mediterráneo, the Royal, the Inglès and the Cala Mayor were building upon the theme of modernism and high levels of service that Adan Diehl had begun at Formentor in the north (Cirer, 2009: 237–248). They expanded their capacities quite considerably: the Mediterráneo added 40 bedrooms and the Alhambra more than 50 between 1928 and 1934. By that year, Palma's 32 hotels had capacity for 1350 visitors while the 46 pensions had 700 (Cirer, 2009: 273). Together these new hotels and the seaside urbanisations founded in the 1920s and 1930s were certainly significant, but the geographical focus remained very much on the urban south coast. For example, the number of hotels, pensions etc. in the *part forana* (i.e. outside Palma) increased by only 21% in the period 1930–1934 while Palma's total increased by 50% (calculated from Barceló, 1966: 53). Note that there was very little hotel accommodation at this date in the two zones that were to be at the heart of the post-war expansion, Platja de Palma and Calvià.

In a detailed analysis of the 1930s, Cirer has recently described this as being a 'boom' in tourism in the years leading up to the Civil War, largely fuelled by the growing affluence amongst those from the Peninsula and especially those from neighbouring Catalunya, with the vast majority using the improved ferry services from Barcelona and Valencia. In addition, the cruise ship industry was beginning to bring considerable numbers to Palma's harbour. By 1933, the income generated by tourism amounted to some 13 million pesetas compared with agriculture's 87 million pesetas. The manufacturing industry still employed almost half of the island's working population. In quantitative and spatial terms, Mallorca remained at some remove from mass tourism, a phenomenon that would develop only after the Civil War and the Second World War (Cirer, 2009: chap. 10). Indeed, there is some evidence that under the Second Republic, national government support appeared not to be wholly in favour of tourism when it introduced laws in 1934 that governed foreign visitors and planning legislation that restricted building on the coast. This led to a wave of protest in Mallorca, including one demonstration attended by more than 10,000 people, addressed by the influential mayor of Palma, Emili Darder, demanding their repeal (Serra & Company, 2000: 76).

Often somewhat neglected by historians of Mediterranean tourism has been the role of the cruise ship industry of the 1920s and 1930s. Percy Waxman mentions various ways of reaching Mallorca by sea in the 1930s, including liners direct from New York (Waxman, 1933: xiv). Mulet (1945) has some useful data from the mid-1930s. For the year August 1935 to July 1936, the number of passenger ships docking at Palma was 360, a tonnage of almost 2.5 million tons and which temporarily disembarked 5300 passengers onto the island. Of these ships, 40% were British, 28% French, 13% German and 11% American. Besides the new 'modernist' cruise liners of the period, many were freighters that carried some passengers. Apart from the American ships, most were part of their respective empire's trade between the home country and their colonies. One ship, the 'Orford', tied up on 1 November 1935, disembarking 42 passengers, taking on seven but releasing 658 tourists for excursions on the island. It was these one-day or two-day visits that first exposed the visitors to the island's delights that, it is suggested, were later translated into longer visits. These excursions were *'en caravana mediante automoviles'*, taking in Valldemossa, Miramar, Deià, Sóller, the caves near Porto Cristo and Artà, Pollença and Formentor. Mostly, they would be in small three- or five-passenger cars, organised by language, rarely on buses or charabancs. Of course, the majority probably never ventured much beyond the confines of Palma's historic core. Mulet goes on to calculate the benefit of these excursionists to the island economy: in 1935–1936, such tourists plus ships' crews spent 3.5 million pesetas, about 14% of the 25.0 million pesetas spent by the total of 430,000 tourists that year (Mulet, 1945: 12). This form of tourism was not to increase significantly again until the early 21st century; in 2006, 649 cruise ships visited the island bringing more than 1 million temporary visitors (see Chapter 8). Meanwhile, the island's tourism industry was to be devastated by war, civil and international.

The Civil War and the Second World War

If new hotels and urbanisations that were built between the turn of the century and the 1930s began to provide the accommodation infrastructure, there remained the question of improving accessibility between Mallorca and the 'continent'. The vast majority of visitors continued to arrive by sea, mostly with Transmediterránia, the principal line founded in 1917, the result of the merger of three companies (Pujalte Vilanova, 2002: 14). However, the 1930s was the decade of the airplane, and its impact on the island's tourism industry was to be profound as we shall see. Comabella and Co. was the first business to establish flights between Barcelona and Palma via seaplanes in 1922. A French company, Latecoere, established service between Marseilles and Algeria via Mallorca. Spain suffered from a shortage of pilots; therefore, many foreign airmen had to be used, but this shortage did stimulate local training, a seaplane pilot school being established at Porto Colom. The Second Republic had established LAPE (*Linies Aeries Postals Espanyoles*) in 1931, another of a number of early airlines linking Mallorca to Barcelona and Valencia including seaplane flights to Jonquets in the Bay of Palma and the Bay of Pollença (Caro Mesquida, 2002: 38). The number of passengers carried on each flight was very limited. The airfield at Son Bonet belonged to a flying club and could, in any case, handle very few aircrafts. The event that was to expose Mallorca to the significance of the airplane was the Civil War.

For the early tourism industry, the Spanish Civil War (1936–1939) was a disaster, although the same cannot be said for the island economy as a whole. Mallorca declared for the Nationalists although there was considerable local opposition to the Francoist side. Many of those who had become successful capitalists, partly through tourism but mostly through the more traditional avenues of industry and banking, gave considerable succour to the fascists. There was only one futile attempt by the Republic to conquer Mallorca with an easily defeated landing at Porto Cristo. Thereafter, the main function of the island was to act as an aircraft carrier for sections of the Italian airforce to bomb Barcelona and other republican strongholds. In addition, the island's manufacturing industries, which had been growing and modernising throughout the 1930s, were turned over to the production of war *materiel*, including military uniforms and boots and shoes. Flows of tourists dried up almost completely despite the alleged neutrality of many of the source countries. Many non-Spanish nationals were advised to leave the island for the duration, including such prominent persons as Robert Graves who had settled in Deià in the late 1920s.

However, the principal effect of the civil war was to be the post-war isolation of Spain by many of its potential trading partners and Franco's policy of self-sufficiency sometimes referred to as 'autarky'. Links within the increasingly 'closed' Spanish state were improved, however, including the establishment of the new air services between Mallorca and the Peninsula (Buades, 2004: 93). The number of passengers coming to Mallorca fell to a low of 7500 in 1940 but by the end of the decade had increased 10-fold to nearly 75,000. Mallorca was not perceived by Madrid to be of much significance to Spain and in any case, its dissident elements had to be purged. Many were imprisoned and often put to work building the island's roads. Political power became concentrated in fewer non-democratic hands, and the economic oligarchy that emerged in the 1920s and 1930s now had even more influence on the island's affairs.

The Second World War saw Spain as a neutral country but with considerable sympathy for the Axis cause. The economy was deliberately isolated by the Western powers who were anxious that Spain should not join the military conflict or offer support to the German/Italian navies operating in the Mediterranean and the Atlantic; the Balearic Islands could have become an ideal base for attacks on Allied shipping. The effect of this isolation was to cripple the Spanish and Balearic economies, especially leading to vastly reduced imports of fuel and food. Efforts were made to increase Mallorca's own food production, but they came to little and by the mid-1940s the island faced a situation close to starvation. One side-effect of the links with Germany and Italy at this time was the opening up of new air routes between these countries and Mallorca as mentioned above, a fact that was to become more significant later in the century, but in 1943 airborne passengers from all sources fell to a low of 2500 although by 1950 this figure had risen to nearly 25,000.

Post-War Mallorca: The Beginnings of Mass Tourism

The development of the Mallorcan tourism industry in the post-war period has to be seen in the context of national politics and economic development, especially in the Franco era to 1975. One of its major fears was that resurgent regionalism at the periphery might undermine the authority of Madrid and lead to what was perceived as the political anarchy that partly gave rise to the Civil War. In fact, both the fear and the centralism of the state have been somewhat exaggerated. In any case, those involved in tourism had a rather different view from the military-industrial cadres at the centre (Pack, 2006(a): 107). Throughout the 1940s, tourism rarely figured on the Franco government's national agenda. Any post-war recovery was to be achieved, it was said, through the medium of industrial development. There was a strong suspicion that overseas tourism could ameliorate the country's economic difficulties only temporarily and that the tourists themselves would undermine the Catholic conservatism that lay at the heart of the state (Pack, 2006(a): 120). There existed, however, a tension between the isolationist centre of the government in Madrid and the more liberal views often held at the periphery. Certainly, there was pressure from the north European countries whose people were increasingly seeking holidays abroad. Because Spain was diplomatically ostracised, it was unable to benefit from the Marshall Plan's European Travel Commission and UK's bilateral tourism negotiations that opened in February 1948. Spain insisted that if foreigners wished to travel to Spain, it had to be under the auspices of Spanish travel agents; at the same time, the British Labour government restricted foreign currency allowances to £25. Nonetheless, the British Workers Travel Association included Mallorca as early as 1949 in its programme (Barton, 2005: 199). Mallorca had already seen such benefits as accrued from tourism in the 1920s and 1930s and had invested in its ongoing development. Despite being generally a supporter of the Franco government, there were those of a more liberal persuasion who saw a healthy tourism industry as a way out of current economic difficulties.

Mulet writing in 1945 began to foresee the rise of tourism on a larger scale than during pre-war provided certain criteria could be met. These included marketing, improvements to internal roads and external communications, rectifying the shortage of hotels, especially high-class ones, and improving cultural facilities. By 1945, 116,900 tourists came to Mallorca, staying for 421,400 bed-nights; honeymooners were a particular target. Mulet and the business community he represented probably saw the future increase in tourism as being a continuation of the same type but on a slightly larger scale of the kind experienced pre-war (Mulet, 1945). With hindsight we know how wrong he was.

Defining when tourism reaches a critical point so that it might be termed a *mass* experience is not easy. Is it a function of number of visitors? Is it the point at which an economy becomes dominated by its tertiary sector's dependence on tourism? These kind of quantitative measures were certainly used by early academic writers on the subject (Burkhart & Medlik, 1974). Others such as John Urry have turned to rather more qualitative aspects such as the Fordist model that focuses on rising disposable incomes in the 'sending' countries inducing an increase in the provision of services in the receiving areas, leading to a circular and cumulative standardised pattern of production and consumption. While recognising that the move to *mass* tourism in Mallorca was sudden and abrupt when examined over a century of growth, we support Joan Buades' assertion that many of the prerequisites were laid down in the 1950s even though the movement may have established itself only in the next decade. He identifies four principal factors in post-war Europe that encouraged an increase in demand (Buades, 2004: 135–137). Firstly, was the post-war reconstruction of the European economies under the Marshall Plan that might include the 'German miracle', the beginnings of the European Community (ECSC 1954/Treaty of Rome 1957) and the nationalisation of industries in countries such as the United Kingdom, leading to reductions in unemployment. Secondly, as post-war incomes rose and holiday times increased, more and more west European people sought to dispose of their leisure times in more distant parts such as the

Mediterranean (Bramwell, 2004: 7). Thirdly, there was through the 1950s, 1960s and 1970s an increase in urbanisation with a greater and greater proportion of national populations living in cities and towns. Leisure space had to be elsewhere. Lastly, was the growth of the airline industry.

This last factor was of the utmost significance. Buades notes that at the end of the war, there were 88,000 aircraft in service worldwide manned by over a quarter of a million pilots, many of which had to find an outlet in peacetime conditions. British tour operators were amongst the most active in acquiring war surplus cargo and troop-carrying aircrafts and converting them to civilian use, often, because most were unpressurised, at some discomfort to passengers. Journeys to the Mediterranean were often achieved only at low altitude and often in a series of 'hops'. Victor Cobb of Thomson Holidays recalls DC3s with bullet holes. Flights to Mallorca took seven hours with refuelling stops. The unpressurised aircraft were very uncomfortable with air sickness a common occurrence (Akhtar & Humphries, 2000: 107–108; Cobb, 2002: 7). Notwithstanding, if any technological device invented mass tourism it was the aircraft (Buswell, 1996: 314).

A key variable in explaining the move to mass tourism is the rate of response, that is, the speed and intensity with which the sending countries could increase their flows of tourists and the extent to which places such as Mallorca could provide facilities such as hotel accom- modation. The assumption is, of course, that the environmental condi- tions exist for the attraction of such tourists. Bramwell (2004), Williams (1996), Fernández Fúster (1991) and many other writers have identified the development of the travel agent/tour operator sector of supply as being crucial in delivering both tourist numbers to Mallorca and in galvanising Mallorcan entrepreneurs into meeting demand in the form of infrastructure, especially cheap hotels.

A second area worthy of further research is the role of the respective governments in the supply chain, including the Spanish government under Franco and various British and German governments in the period 1945–1960. It would be equally interesting to discover the role of the governments of wartime neutral counties such as Sweden and Switzerland. An examination of the intertwining of tour operators' interests and those of the two sets of governments, especially at the civil servant level, would pay fascinating dividends.

The role of the tour operator is thought to have been crucial, but, as Towner has pointed out, the archives of tour firms are far from complete and not entirely reliable for narrating the growth of mass tourism (Towner, 1996: 253).

Organisations such as the Workers Travel Association in Great Britain had included Mallorca in its itineraries during the 1930s but by 1949 it was to reappear. Its publicity stressed that the 'Island of Charm' had much to offer working-class holidaymakers, including beautiful scenery and quaint folkways such as gypsies and flamenco, the latter two hardly part of the island's culture (Barton, 2005: 199). Other researchers have turned to the biographies of the principal actors in the tour operator business in which the arcane processes whereby visitors were transferred to Mallorca and accommodated for their fortnight's holiday are often amusingly recounted. Amongst British firms two are worth recalling: Horizon Holidays (founder Vladimir Raitz) and Pontinental (founder Fred Pontin), both of whom became important in Mallorca.

Raitz's first location was Corsica, but he then turned his attention to Mallorca where,

> Upon arrival, Majorca made a gentle impression. Certainly it was an attractive island, with high mountains and deserted beaches. There was a charming little airport with a terminal building resembling a country house... the island offered great potential for tourism... One problem was the lack of hotels. All of them were of pre-war vintage, and the large comfortable ones could be counted on the fingers of one hand. (Bray & Raitz, 2001: 4)

The British government throughout the second half of the 1940s decade was constrained by its need to rebuild its economy after the war, and the economic emphasis was put on exports, which meant strict currency restriction and limits on foreign travel. The amount of money that a British tourist could take abroad was limited to £25, rising to £100 by 1955. All sorts of ruses were resorted to to enable people to travel abroad, the most ingenious of which was the 'educational visit', which for cultural reasons had official blessing. Inventive travel agents enabled many so-called choirs, sports clubs and women's groups to take advantage of this fictional designation. The 1947 Civil Aviation Act in the United Kingdom was drawn up to protect national (-ised) airlines such as BEA and BOAC from charter companies who could hire out planes for as little as 2/6 d (25 p) per mile (Barton, 2005: 200). By 1951, the new Tory government was willing to relax controls on foreign travel to a greater degree and the new Air Transport Advisory Council abolished the student/teacher clause and liberalised the issuing of licences. Raitz recalled that

> As soon as government liberalization policy had been announced I made a quick trip to Majorca to look at a hotel in the NW corner that had been recommended to me (RJB – actually in Port de Sóller)... I made a provisional contract with the owner of the hotel... Not all our clients found Noreen's (the hotel owner) exuberance and salty language exactly their cup of tea. (Bray & Raitz, 2001: 17)

By 1957, Eagle Aviation, a rival firm, offered a 15-day package to Mallorca for 38 guineas with 'all the magic of swift comfortable travel in gleaming Viking planes' (Bray & Raitz, 2001: 50), but early package holidays could not be less than the airfare unless it was for closed or affinity groups such as civil servants or the infamous West Hartlepool Bird Fanciers Circle!

Fred Pontin had been a pioneer in the post-war British holiday camp business (Ward & Hardy, 1986) but by the early 1960s was turning his attention towards Mediterranean opportunities including Mallorca. As he noted, '. . . This was a time of cheap charter flights . . . land was cheap, building costs, providing they were carefully controlled, were comparatively low and catering costs were competitive' (Pontin, 1991: 82). This meant that holidaymakers could be offered two weeks on the island for the same price as a holiday at one of his British camps but with all the advantages of the Mediterranean summer plus 'cheap booze and plenty of it'. It would appear that many of these early British entrepreneurs worked closely with certain sectors of the government and, according to Raitz, were able to influence policy in their favour.

To understand the role of the Spanish government of this period it is important to set it in the context of world affairs. The post-war period soon degenerated into the Cold War, with the United States leading a crusade against what it saw as a Soviet-led movement for world domination. Amongst the first manifestations of this were the Korean War (1950–1951), the Malayan insurgency (1948–1960) and disturbances in the Dutch East Indies (1945–1949). Successive American governments under Truman, Eisenhower and Kennedy sought, where they could, to bolster anti-communist regimes, including Spain's fascist government. This included the deployment of an American fleet in the Mediterranean whose ships and sailors were able to take full advantage of Mallorca's facilities for rest and recreation (R and R) (Buades, 2004: 121). It also meant offering economic support in the form of trade, especially in industrial raw materials, foodstuffs and energy resources in return for military bases. At the same time, it was vital for Spain to increase its export earnings, which throughout the 1950s were in deficit. Franco's original notion of a self-contained Spanish state did not live up to expectations. The experience was to the contrary, namely, that it simply made Spain poorer; average incomes fell and income disparities widened in favour of a narrow elite. Tourism was touted as a means of increasing foreign exchange, essential for trading purposes since the peseta was a non-convertible currency. A key to the opening up of the Spanish economy in the 1950s was an American credit guarantee of $62 million designed to lead to more inward investment to support Spain's currency. Unsurprisingly, large-scale inflation soon followed accompanied by devaluation, which quickly made Spain an even more attractive destination. Tour operators and overseas holiday companies soon found that their harder currencies placed them in a strong bargaining position with hotel companies, and encouraged inward investment (Alzina *et al.*, 1992: 390). Progress along this path in the Spain of the 1950s was slow, with tourist numbers increasing from 1.26 million in 1951 to 4.19 million in 1959. The first 'boom' on the road to mass tourism in the country may have begun in the late 1950s, but the real take-off was to occur in the 1960s; even as early as 1963, numbers going to Spain had already risen to 8.7 million. Three questions arise: How did the Spanish government intervene in the economy? What happened between the mid-1950s and the early 1960s to accelerate the beginnings of mass tourism that we have described? And thirdly, what did this mean for Mallorca?

Spain may have thought itself a totalitarian state, but in practice much of its management was chaotic and it relied on the private sector for its economic development in the post-war period. As with most government policies that wished to encourage the private sector in a particular direction, Spain advocated a 'stick and carrot' approach. The 'carrot' consisted of various fiscal measures to encourage inward investment, in effect, lowering the cost of production. These could take the form of tax concessions, discounts or offsets and access to cheap credit. Another area of government intervention was by providing infrastructural support – building roads, water and sewage systems, street lighting, airport construction etc. The 'stick' was rarely used in practice; its principal medium was planning control, limiting developments in some areas and encouraging it in others. Indeed, the weak planning system proved to be more of a 'carrot', enabling certain locations to be granted planning permission with little opposition. Another 'stick' could have been to place environmental restrictions on developments, but these were largely ignored, laying the foundations for the negative effects of inappropriate development that were to emerge 20 years later in the 1980s. However, as Newton points out, although the centralist government kept a 'tight and unimaginative grip' on tourism with an emphasis on mass tourism on the coasts in order to maximise foreign exchange, despite their claims the Francoists' controls over many aspects were minimal and subject to much corruption (Newton, 1996: 142).

The principal means of encouraging the development of the tourism industry was the use of bank credit for hotel building. The Crédito Hotelero introduced as early as 1942 underwrote loans for up to 60% of the total building cost of any hotel. The loan would then be extended by commercial banks at favourable rates of interest. Introducing a new cadre of technocrats, especially economists, into government in the late 1950s resulted in a very necessary liberalisation of Spain's economy; foreign currency reserves had fallen to an all time low of $6 million by 1959. Following the French example, the approval of a Stabilisation Plan in

1959 began a long process of economic liberalisation. Its main aim was, firstly, to restrict demand and slow inflation by introducing tough fiscal measures and, secondly, to open up trade to foreign companies and encourage inward investment. The plan's initial deflationary effects saw a rise in unemployment and the beginnings of the great northward migration of Spanish workers to the growing economies of Germany, France and the Netherlands, but slowly its benefits were felt and foreign tourist numbers rose. By the end of the 1960s, more than 20 million tourists accounted for more than 9% of Spain's gross national product (see Figure 3.2). Overseas remittances also brought in hard currencies. Overseas businesses, especially American, German and French, now felt Spain to be a country worth investing in. For tourism this led to the beginning of the first real 'boom'.

In Mallorca its effects were astounding. Salas Colom (1992: 70) records that there were only 105 recognised hotels and hostals in Mallorca in1950; by 1959, there were 443. This decade was a transitional one in which hotels in the higher categories were still dominant; the third class that was to become the principal type in the near future was only slowly emerging; it was to remain the staple from then on. Most of the growing numbers of tourists found themselves in older hotels that were expanded or in the many pensions that provided additional accommodation. In the early 1950s, both expansion of existing stock and new build retained, architecturally speaking, fairly strong allegiance to Mallorcan styles but as pressures grew and the nature of tourism changed towards a mass market, building techniques and styles became more rational and international. The great 'boom' in hotel building of the 1960s was heralded by new regulations from the central government dating from 1957 and 1960 when very precise measurements for bedrooms, dining

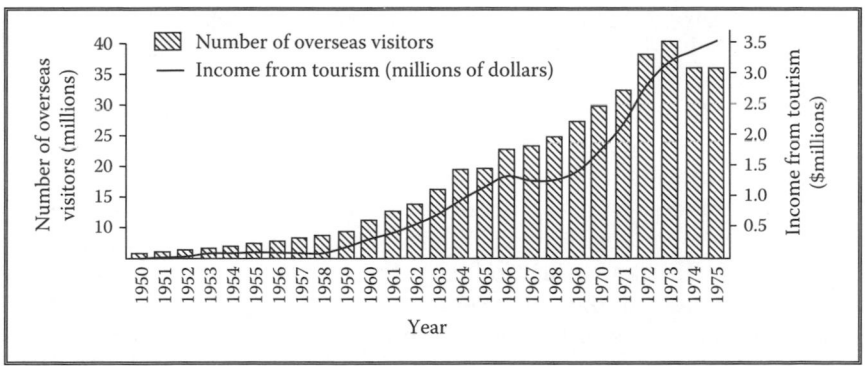

Figure 3.2 Growth of overseas visitors and income in Spain (1950–1975)
Source: Barciela, López, Malgarejo & Miranda, La España de Franco (1939–1975)

rooms and public rooms were recommended that would eventually lead to the new classifications (Seguí Aznar, 2001: 104). In the 'boom' itself, these regulations were often to be honoured only in the breach.

Such hotels were themselves a physical expression of the new Spain and Mallorca, but in effect this meant that they had to reflect the almost industrial nature of the tourism experience under 'mass' conditions. This Fordist approach led to large units of identical rooms, all the same size, all with same equipment, built to accommodate the large numbers of tourists being increasingly decanted from the aircraft at Son Bonet. Only the de luxe hotels, of which there were only two in Mallorca in 1950, could offer choice as an element of accommodation. Designed by a new modernist school of architects, stylistically they reflected this rational approach rather than any regional historicism (Seguí Aznar, 2001: 113), part of the new urban landscapes of the coastal settlements. Compare, for example, Hotel Maricel in Cas Català (1948) and Hotel Bendinat (1951) with the Nixe in Calamayor and Fenix of 1957. With rooms built on modular lines, these hotels adopted a proper orientation towards the sea, with individual balconies optimising their exposure to the sun. The ground floor public rooms would open out onto private gardens and swimming pools (Seguí Aznar, 2001: 119). The design, the building materials, the layout and the settings spoke of an expectation of luxury to suit a middle-class market. What transpired as the first boom took off in the 1960s was rather different as the 'new' tourists had rather less money to spend on accommodation thanks to the tour operators' keen price competition. The aspirations for the luxurious hotel soon gave way to the rather more functional three-star hotel for the tourist masses, usually located in the new complexes in the rapidly urbanising coastline. This was the beginning of the 'wall of cement' exemplified on the east coast at Cala Millor, Sa Coma, S'Illot and Cales de Mallorca and on the south coast in what became known as Platja de Palma.

Finally, as we enter this new era, it is worth asking, as does Manera (2001: 320), 'why did mass tourism erupt in particular in Mallorca in this period'? After all, many other Mediterranean islands such as Corsica, Sardinia, Sicily and Malta had a similar resources base for tourism and we have seen that north European entrepreneurs were in the late 1940s and 1950s exploring their potential too. Following his general theme was there something special about the Mallorcan economic system, namely, its experience of adaptation to changing circumstances? Manera cites the 18th-century wine and spirit industry, the development of the shoe and textile industries in the next century and the agricultural and food processing industries of the early 20th century as examples of business innovation in response to changing patterns of demand. Tourism displayed from the late 1800s another element of this combination of endogenous entrepreneurial spirit and external stimulus and linkage, a

Search

Mass tourism

Fishing

4. - changes in tourism demand → environmentally conscious tourist
'New Tourist'
Q: include or focus more on Mayorca? Meet demand — change tourism product

5. - Env & SD

6. Env & Sou Intervention — positive impacts - A MOVE layout
full course - CIA Benefit → mainstyle tourism
- including examples positive

7. Conservation in Mayorca

- Describe impact - briefly → Analyse // HOWEVER!! [
- why & how occurred → Issues
→ type of tourism allowed that tourism to thrive - needs to be managed

250-300: Intro - Conclusion ✓ / Environmental impact less
- Developed as a mass tourism destination ↓ whittle down 3
- Factors & reasons

1. Paras/paragraphs - Short with description
2. One of the negative impacts of Mayorca is because ... according to ...

Questions

1. Rant about Terry
   - Discrepiancy's
   - Discrepiancy's in Marking
   - He & another student had to exact same coments yet she got a higher grade & didn't put the proposal in the template he requested

   - asked for more feedback didn't get any
   - Assignment said examples not stated is feedback! had to ask

2. Environmental impacts
   - Should I include socio-cultural impacts $ as a result of environmental impacts or focus on environment
     e.g. scarce resources
   - should I include scarce resources
     e.g. scarce resources

   - effects on people
   - stick to environment

* MAJORCA - Named destination

3. Layout
   1 - Overdevelopment due to lack of planning
      - aesthetic pollution
   2 - Pollution - waste, water

   - negative

Mix in 1 neg & 1 positi ~~~e~~~

Verry & Larsen (2011)

feature that many other Mediterranean island economies did not have. Mallorca was not isolated from wider economic forces or ideas as so many have previously believed and where Mallorca led other coastal and island regions were to follow. In this sense, Mallorca and to a lesser extent Ibiza were to become laboratories for Mediterranean tourism (Seguí Llínas, 1995, 2006). Growth in the era prior to the Civil War and the Second World War reflected an evolving global tourism particularly in the cruise liner sector, but it is difficult to see mass tourism's beginnings in the 1950s as simply a continuation of past trends as Cirer does (Cirer, 2009). The next chapter will attempt to isolate the factors that contributed to a massive step-like change that occurred in the 1960s when tourist numbers to Mallorca increased from 400,000 at the beginning of the '50s decade to more than 3.5 million 13 years later.

Chapter 4
When Majorca Was Spelled with a 'J': The First 'Boom' of the 1960s

Introduction: Towards the First Boom

In many ways, the growth of the tourism industry in Mallorca until the late 1950s had been significant but not spectacular, building on the foundations laid in the 1930s but interrupted by two wars (Shor, 1957). Buades has shown, however, that the post–Second World War period in Mallorca should not be underestimated in its contribution to the first and best known 'first boom' (Buades, 2004: chap. 4). The old Mallorca airport of Son Bonet, for example, processed about 75,000 passengers in 1950, a figure that had risen to only a quarter of a million by the middle of the decade. In 1950, 98,000 tourists had stayed in hotels in the Balearics. This figure increased fourfold in the following decade. Bed spaces rose from more than 610,000 to 3.75 million. Initially, two thirds of the visitors were Spanish and some French; only about 200 German visitors and less than 3000 British visitors were staying in hotels in 1950. But things were about to change dramatically (Figure 4.1). The following rather bald figures give some indication of the scale of growth. From 1960 to 1973, the number of visitors increased from 400,000 to 3.6 million, a rate of increase of 60% per annum, that is, an average of more than a quarter of a million each year (Table 4.1). The number of bed nights rose from 4.9 million in 1960 to 54.2 million in 1973, and more than 140,000 hotel places were built, with the industry directly creating 100,000 jobs (Buswell, 1996: 311).

This was truly the beginning of the first 'boom'. In Mallorca, its effects were astounding. By 1960, British visitors to the Balearic Islands had overtaken Spanish ones but German numbers were still low at about 37,000, the same as from Sweden (n.b. official figures for individual islands were not recorded until much later) (Ministerio de informacion y turismo, 1965: various tables). In 1964, 136 hotels and pensions were added to the Balearic stock, 131 were expanded and seven closed down, resulting in a net gain of 129 establishments, 5930 rooms and 10,400 bed spaces in one year alone. By the end of 1964, Mallorca had 918 hotels and pensions, 27,286 rooms and 48,908 bed spaces. In a series of curious comments, the official statistics for 1965 noted amongst other things that in 1964 alone a new hotel was opened on the island every two days, four hours and 48 minutes, that if all the tourist beds in the Balearic Islands were laid end to end they would stretch from Sevilla to Cordoba or 15 times the height of Everest and Kilimanjaro added together and that

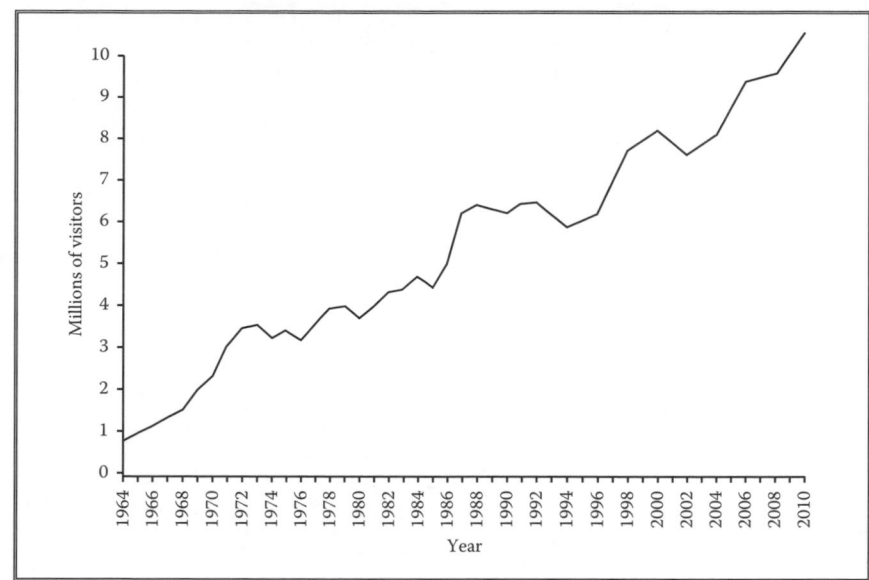

Figure 4.1 Visitors, 1964–2010
Source: Conselleria de Turisme

there were two tourists for every inhabitant. And hotels were not the only additions to the landscape in this era; for example, 62 new restaurants seating more than 4000 people were opened in Mallorca in 1968. By the end of this period, the spatial pattern of tourism in Mallorca had been established. Virtually no new resorts were built after the mid-1970s but many were expanded, consolidating the locational patterns of this period.

Growth Factors 1960–1973: The British Are Coming!

The growth of tourism in the Mediterranean in the late 1950s was not as spatially widespread as might be expected. British travel agents, tour operators and airlines tried many of the less developed areas of the Mediterranean, including Corsica, Sardinia, Malta and Cyprus as well as parts of the North African coast, to begin developing holidays besides Mallorca. It is not unreasonable to ask why Mallorca was so successful at this early stage when other places were not? We have noted in the previous chapter that Manera has tried to make a case for certain factors in Mallorca that did not exist elsewhere (Manera, 2001: 319). In examining these factors, the point made above about the exogenous elements derived from west European travel industry activity must also be taken into account. Mass tourism in Mallorca cannot be explained by indigenous factors alone. In September 2010, Mallorca's most important

Table 4.1 Visitors to Balearic Islands, 1950–1984

	1950	1955	1960	1965	1970	1975	1980	1984
United Kingdom	2,680	32,756	93,335	261,010	757,087	1,086,918	1,024,473	1,718,184
West Germany	229	15,245	37,153	196,452	448,872	778,938	939,196	1,177,372
Spain	66,525	57,920	86,688	176,338	243,767	466,137	447,667	552,558
France	18,576	33,917	78,561	102,482	171,491	249,793	325,754	359,512
Sweden	585	4,326	37,726	61,199	107,223	129,011	151,388	76,741
United States	870	10,790	23,484	74,909	105,625	40,477	23,484	74,909
Total	98,081	188,704	400,029	1,080,826	2,274,137	3,435,799	3,550,639	4,737,279
Bed nights	610,500	1,521,751	3,776,525	11,232,296	26,128,563	36,249,765	33,317,732	50,161,965

*All Balearics, only those staying in hotels.
Source: Ministerio de informacion y turismo (1965): El turismo in Baleares (1965). Delgacion provincial de Baleares and Dades Informatives (various years)

local newspaper, the *Diario de Mallorca,* recorded the death of Vladimir Raitz, the founder of Horizon Holidays, in a brief 50-word note. British newspapers *The Guardian* and *The Daily Telegraph* honoured him with whole-page obituaries. Therein lies a difference of historical perception. To Mallorcan writers such as Cirer, mass tourism was largely invented in Mallorca, but it can be better argued that a much greater contribution was made by British entrepreneurs such as Raitz – an example of demand triumphing over supply (*Diario de Mallorca,* 17 September 2010; *The Guardian,* 9 September 2010; *The Daily Telegraph,* 3 September 2010; Cirer, 2009). It is useful to compare, say, Corsica where Raitz's first venture took place, with Mallorca; the former had little tourist infrastructure in the 1940s and 1950s, whereas Mallorca had sufficient to begin to attract overseas investment to be grafted onto local energies.

In the past it has been assumed that mass tourism was an economic response to the post–Civil War poverty of Mallorca, but Manera and Garau-Taberner are at pains to point out that its economy was not in crisis and, in fact, already had in place not only some of the requisite tourist infrastructures set up in the 1930s but also a noteworthy tertiary sector (banking, transport, travel agents, small hotels, specific resorts etc.) to support it. In the 1950s, income per head was already about 20% above the Spanish average and was even about three quarters of the European average (Manera & Garau-Taberner, 2009: 32). On the other hand, there were short-term adjustment problems such as serious food shortages and while agriculture and industry may have been losing employment, output at this stage was kept up through technological improvements. Historically, the island had always had an open economy; importing tourists might be seen as an historical continuity (Manera & Garau-Taberner, 2009: 43). They also believe that flexibility in the labour force contributed to the rapid growth as workers moved from the fields and the factories to the hotels and bars, but surely this can have been achieved only with a certain *force majeur.* We shall now show how there were frictions in the labour market and above all shortages that only inward migration could compensate for.

Picornell and Seguí (1989: 78) have attributed the growth of tourism in this period to the continuing development of the principal factors identified earlier: the improving quality of life in west Europe; the creation of charter airlines and the rise of the tour operator; the modernisation of the Spanish economy; inward migration of population and workers and the rising importance of Palma, Mallorca's capital city. In other words, the influential variables may be seen as continuing into the new decade but their combined effect now was to stimulate a massive increase in the scale of operations. Scale appears to be the most important aspect of change but were new innovations introduced in order to accommodate it?

The rising affluence within west European society meant larger disposable incomes and a reduction in the length of the working week. The significance in the United Kingdom of the holidays with pay legislation introduced in 1939 did not really become effective for overseas holidays until after the war (Kynaston, 2007: 217). By 1961, 3.5 million British people holidayed abroad, compared with half that number a decade earlier (Hennessy, 2007: 539). It was also the era of the West German 'wirtshaftwunder' though its impact was not really felt until the next decade in Mallorca when years of austerity pent up under rigid post-war conditions were released onto overseas markets.

Secondly, the transformation of the transport infrastructure for holidaymakers travelling from north and west Europe to the Mediterranean was highly significant. Important here was the substitution of land and sea transport with air travel. Again, we saw the origins of this in the 1940s and early 1950s in the previous chapter, but from about the mid-1950s the scale changed markedly. Four factors were important: the technology of the aircraft themselves, the rise in the number of independent airlines, the construction and expansion of airports and the innovations within the package holiday system so that relative prices made air travel accessible to a mass market.

In aircraft design and performance, the most important move was towards turbo-prop and jet engine propulsion (Lyth, 2003, 2009). This had two benefits: the ability to fly at higher altitude, which reduced costs per mile, and much increased speed so that travel time to the Balearics from Germany and the United Kingdom was reduced to three and soon to two hours. It has been estimated that the cost of charter flights between London and Palma fell 27% in the period 1967–1971 (Lyth, 2009: 18). In the early 1960s, most of the major airlines' energies were concentrated on long-distance journeys from capital city airports, particularly between Europe and North America. For Mediterranean tourism what was required was a series of companies devoted to this short-haul market operating outside the large national airlines. For the Mallorcan market from the United Kingdom the response came from airlines such as Britannia, Dan Air and British Midland. To expand geographical markets for tourism in southern Europe, investment in provincial airports close to regional markets was needed. The political and administrative geography of post-war West Germany was deliberately constructed on a regional basis and so the evolution of regional airports in the 1970s appears as a logical development, but Britain was much more strongly centralised. The effort to develop local airports was a remarkable achievement that was to benefit overseas tourism for the working classes. It most cases, it was done by expanding wartime airfields, many of which had actually begun as civilian airfields in the 1920s and 1930s. These were then requisitioned by the Royal Air Force (RAF) during the Second World War, their post-war growth being a combination of local entrepreneurship and cooperation between the private sector and local authorities. Examples include Newcastle (Woolsington) in the north, Bristol (Whitchurch and Lulsgate Bottom) serving south-west England, Cardiff (Rhoose) in South Wales and Edinburgh (Turnhouse) in eastern Scotland. Stansted in East Anglia was one of the earliest post-war developments when London Aero Motor Services bought 18 Halifax bombers and the airfield located there in 1946 and converted it for civilian use (Thomas, 2005: 58). Although some 'intercity' flights were provided, the primary trade was for the tourist market, especially to the Mediterranean. In Germany, the principal airport serving the main nucleus of population, the Rhine–Ruhr region, was Düsseldorf, which was opened in 1927 but expanded considerably in the 1960s when investment was undertaken by the Länder government. This airport was to develop as the home of Germany's largest charter and travel company Lufttransport-Unternehmen GmbH (LTU).

If the airports of West Germany and the United Kingdom were important, even more so was the provision of new facilities in Mallorca itself. The original airport serving the island was Son Bonet to the north of Palma, originally a private flying club field converted for military use during the Civil War when Italian and German aircraft were based there. It became the official Palma airport in 1947 although Robert Graves and his family had landed there when returning to Mallorca in 1946; by 1950, 75,000 passengers were using Son Bonet (Salas Colom, 1992: 53). However, this airport was too close to the growing industrial district of Pont d'Inca and Palma's suburbs, and another site capable of expansion was required. This was found at the military airfield of Son Sant Joan 8 km to the east where abundant state-owned land was made available. Despite its proximity to the growing tourist settlements such as Coll d'en Rabassa, commercial flights were transferred there in 1959. Three years later, more than 25,000 flights were bringing over 1 million passengers to this new airport and by the mid-1990s it was the 10th largest airport in Europe in terms of passenger throughput. By 2008, more than 9.6 million entered Mallorca through a greatly expanded airport.

As we saw developing in the 1940s, the charter airlines began with second-hand ex-war planes bought cheaply from British and American governments anxious to offload surplus requirements. By the 1960s, a market in new and second-hand jet airliners began to emerge but in order to make these pay the charter airlines and the tour operators had to make them work harder in order to cover costs (Barton, 2005: 207). In many ways, acquiring aircraft was relatively easy; obtaining the necessary permits in post-war Britain to fly them to the Mediterranean proved to be much more difficult. Only by the mid-1950s – a decade after the cessation of hostilities – did it become possible for the true charter airline to have relatively unrestricted access to the skies. The revolution, however, came in the 1960s with the introduction of the jet-engine planes, with many charter airlines shunning British aircraft for the new American Boeing and Douglas models.

Equally significant was the way in which the travel company – and often they were merely travel agents – either directly chartered planes or increasingly, actually owned or leased them. The merger of High Street travel agent with charter aircraft led rapidly to the emergence of the tour operator and the modern package holiday.

Shaw and Williams (1994) analysed this form of business activity some years ago noting its remarkable and long-lasting effect on Mallorca. Barton and others (Barton, 2005; Brendon, 1991; Walton, 2002 etc.) have recorded its origins in Britain in the trips organised to the Great Exhibitions of 1851 and 1864. Its modern form linking west Europe to the Mediterranean was started from small beginnings by far-sighted single entrepreneurs who detected opportunities and capitalised on market openings. Victor Cobb, for example, began life with a small Blackpool travel agency – Gaytours – eventually rising to become managing director of Thomson Holidays. He recalled that in 1960 Mallorca had little by way of beach-type accommodation; Palmanova and Magaluf were largely deserted beaches, while S'Arenal had only two hotels (Cobb, 2002: 37). When Skytours 'discovered' Cala Millor, there was little or nothing there, but by 1964 it had many new, up-to-date hotels (Photo 4.1). Thomson Holidays purchased Arenal Park Hotel and leased four more, including the Honolulu in Palmanova. Thomson eventually owned 17 hotels in the Mediterranean, with 12,000 beds, employing 2500 staff, setting a benchmark for quality control (Cobb, 2002: 170).

British holiday camp operator Fred Pontin also saw new opportunities for his type of operation in Mallorca, opening his first camp at Cala Mesquida (Artà) in 1964, 1 of 10 camps in the Mediterranean area (Obituary, *The Daily Telegraph*, 23 August 2001). It was based on his concept of such holidays in the United Kingdom where, by the 1970s, his firm had developed 22 such camps to rival Billy Butlin's empire. Cala Mesquida had 142 brick chalets, a shower bloc and a dining room; it was a self-contained holiday village where a fortnight's holiday including travel cost just £50. By 1976, there were three Pontinental holiday camps in Mallorca, but by the end of the decade their rather austere provision could not compete with hotel-based package holidays (Pontin, 1991; Wright, 2002: 185–186).

Some tour operator firms were to develop into large-scale multi-national concerns. By the 1990s, tour operators accounted for about 25% of world tourism, of which half was provided by European firms (Čavlek, 2000:174). Initially, the tour operator was in essence a marketing organisation that recruited tourists through advertising and other forces

Photo 4.1 Art deco style built in 1960s in Cala Millor

in the supplying country and delivered them to the resorts where the actual holiday was provided by local hotel firms. This is what Čavlek calls the take-off stage. By 1963, some 630,000 UK tourists travelled to Europe and by 1970 out of the 5.75 million who took European holidays about 2 million were being transported by air charter companies. The first German package tours date from 1962 although the Condor airline was operating from 1956. By 1970, 2.8 million German tourists were on package tours. By the late 1950s, some vertical integration of these operations was beginning to emerge as economies of scale were sought so that numbers could rise and costs kept down. Gradually, the once segmented industry began to realise the advantages of closer links between all parts of the system. This eventually led to the integration – and in many cases mergers, acquisitions and takeovers – of travel agents, hotels and charter airlines. This shifted the industry from a producer-led activity to a consumer-based one in which marketing and sales were the dominant features. With an increase in firm size, diversification of product soon became a priority to avoid over-concentration in one geographical area – the Balearic Islands or Mallorca – and so the wide range of destinations now available in today's holiday brochures soon became the norm.

Thirdly, throughout the 1950s, it became increasingly clear to the Spanish authorities that a restructuring of the national economy would be required in order to meet the balance of trade objectives alluded to in the previous chapter. If foreign currencies and export earning were to increase through the medium of tourism, then a more stable economy would be necessary. In essence, this meant making the peseta more attractive by devaluing the currency, in effect by as much as 42% against the US dollar. It was the increase in tourism receipts during the 1960s due to the attractive new exchange rate of the peseta that was to help pay for these. By 1965, tourism receipts as a percentage of total exports reached a remarkable 111% of the value of manufactured goods, covering 64% of Spain's trade balance (Sinclair & Bote Gomez, 1996: 91).

In the tourism sector, investment, both foreign and national, went into building infrastructure: land, hotels, apartments and second homes plus all the structures necessary to support this activity, especially the construction of roads, water and sewage systems and energy supply. Where private capital could be deployed for such utilities it was, as in the case of electricity supply in Mallorca, but the public sector had to fund some of the supporting infrastructure – such as water supply – which at the national level was achieved by an increase in government borrowing; this was not so readily forthcoming. To these supply-side improvements, led by the reform of the national political economy, should be added, for example, electricity supply where the local company, GESA, established by the banker Juan March as early as 1927, had to be expanded considerably. This meant not only increasing the output via a new power station at Alcúdia in the north of the island but also improving electricity distribution by using high-tension cables. Generating capacity rose from about 16,000 KW in 1952 to more than 100,000 KW by the mid-1960s (Casanovas, 2005: 48; Pujalte Vilanova, 2006).

Clearly, tourism was being singled out as the milch cow to underpin these developments for a long time into the future. The question at the time was: were tourism numbers likely to hold up over a long enough period? Holidays, after all, are only positional goods and tastes and fashions can soon change. In the case of Mallorca, and to a lesser extent, in Spain as a whole, the risk proved successful at least until the energy crisis of 1973–1974 and the recessions of the early 1990s and 2007–2010.

For Mallorca, the results of this economic invigoration through tourism were spectacular, largely because of the response of local entrepreneurs and politicians and overseas investors, converting the island into what Buades has called 'un paradís immobilàri' (a real-estate paradise). It was a process that began before 1959 and was often accompanied by many corrupt practices. In some ways, events in Mallorca before the Stabilisation Plan came into being demonstrated what could be achieved by way of tourism. The key variable was the acquisition of land, which meant coastal land, of course, fronting onto beaches, followed by changes in the planning regulations designating such zones for resort development often with permission for investment by foreign interests. 'L'effect ... serà un transformació radical del paisatge balear i del litoral espanyol en general, el qual serà ofert al Moloch d'operacions immobiliàries sense precedents ('The effect would be a radical transformation of the Balearic and Spanish landscape and coastline which was to see greedy real-estate operations without precedent') (Buades, 2004: 142), turning Mallorca into a veritable gold mine for the Spanish government (Barceló, 1961: 141). In the decade from 1950, at least 10.5 million square metres of land in coastal zones was to be reallocated to urban uses, almost half of it in Calvià municipality.

An example of the processes involved in land-use change to tourism in the Bay of Palma is given by Picornell (1986). The area to the east of Palma was to develop as one of the main locations for hotel development in the 1960s and 1970s.The low-lying area behind the coastal dune strip was known as the Prat de Sant Jordi. Attempts at drainage were begun in the early 19th century using dykes and windmills, but the key to agricultural change was the break-up of the main estates such as that of Son Sunyer, which covered 1200 ha including the whole of the Platje de Palma. Beginning in 1900, this estate was divided into 362 small holdings, the vast majority of which were under 5 ha in size. These were mostly bought by the professional middle class from the city as investments, with tenants producing market gardening produce from these *hortas*. The land, including the 5-km beach, to the seaward side of the poor coastal road had little perceived value in this sense. A number of small settlements existed along the coast, including a collection of fishermen's huts at what was to become S'Arenal. The key to tourist development was the creation of an urban zone along the coast aided by improvements in the road system and the opening of the railway between Palma and Santanyí, making it attractive to *palmesanos*. With the opening of the planned settlements of C'an Pastilla and Coll d'en Rabassa in the 1920s and 1930s, the seasonal suburban expansion of the coastal strip began, to which it was relatively easy to graft the new intensive hotel land uses of the 1960s. However, much of the new (1960s) construction was unplanned and with little or no infrastructure (Picornell, 1986: 11). Further expansion into the *hortas* rapidly converted agricultural land into land for hotels, apartments and housing for the immigrant workers who were actually building properties and providing services to tourists. Once S'Arenal became established, the urban strip continued further eastwards with the new settlements of Cala Blava, Bahia Azul, El Dorado etc. Note here once more the continuing use of place names that had no historical roots but which were really the invention of developers. Historically, the coast of Mallorca had rarely been settled because of fear of corsair and pirate attacks, leaving the field clear for the vivid imaginations of estate builders to get to work.

Population in this coastal zone increased from 583 in 1930 to 2560 in 1960 and to 13,427 in 1970 (Picornell, 1986). Sbert Barceló has drawn attention to the key role at this time of the Head of Coasts for the Balearic Islands, the engineer/planner Antonio Garau Mulet. In 1964, he attempted to draw up a plan for the 5-km 'front line' of this area between the outskirts of Palma and S'Arenal that could bring together the legitimate concerns of private landowners on the one hand who had begun to develop the area from the early part of the 20th century and the municipalities of Palma and Llucmajor on the other, the latter two representing the public interest. At the same time, he had to reconcile these two interest groups with the aspirations of the national government for tourism. While he foresaw the obvious economic benefits from the spatial and environmental transformation of the Platja de Palma, the crucial question was: who was to bear the cost (estimated to be 160 million pesetas in 1965 – an enormous sum) – the hoteliers, the municipalities or the national government? Hoteliers and tour operators might construct hotels but who was to be responsible for promenades, public open spaces, berths for *club nauticos*, even the maintenance, and in the case of C'an Pastilla, the creation of the beaches (Sbert Barceló, 2007: 58–60)?

Consideration of a fourth important variable that helps to understand the growth of tourism in this first boom period (1960–1973) is the labour supply situation. By the mid-1950s, unemployment, and more particularly under-employment, remained a major local problem. We saw earlier that through the first part of the 20th century, despite improving industrialisation, job opportunities lagged behind population growth, leading to emigration as a partial solution. At the same time, rationalisation of agriculture – the main employer – witnessed a reduction in land-based jobs from about 40% of all jobs in the primary sector to about 14% by the mid-1970s, stimulating rural to urban migration, especially of young people. The early advocates of tourism saw this shift as a new employment opportunity. The pre-war growth of tourism had certainly laid the foundations of a new cadre of tourism workers in the 'grand' hotels of the period, but by the 1960s the challenge was intensified by the sheer scale of growth. 'Evolution' was forced aside by the need for 'revolution' in the labour market (Mascaró Pons, 1989). Two sectors can be identified: the direct employment in hotels and related activities, and construction. In both, a number of factors caused friction in the labour market, preventing tourism from being the immediate panacea. First and foremost was the mismatch of skills between the rural labour supply and those needed by the tourism industry of the new urbanisations. Perhaps the skill levels of potential bar workers, cleaners, bedmakers and waiters might be seen as relatively low and readily acquired; nonetheless, the culture of work differences for village young men and women was considerable. At a higher skill level, such as management, financial control, marketing and catering, the problem was more acute. Although government investment in higher education and training was eventually to alleviate the shortfalls, most of this would not come until later (Moreno Rodriguez, 1990: 110). Meanwhile, immigration was a partial solution. A second friction would have been that of distance, never great in a small island, but it should be remembered that historically Mallorcans were essentially inland people and tourism was located on the coast. Should workers continue to live at home and commute from the old villages and farms or seek new residences? Thirdly, the new tourism was located in urbanising zones with an initial focus on the Palma area, which meant apartment living, often in multi-occupation, rather than the farm or the village house. In the event, these frictions proved small and short-lived for those seeking work in the burgeoning bars, restaurants and above all, hotels.

In the construction industry (and the related raw materials industries such as quarrying, cement-making etc.), the friction was greater. Again, it was the pace of change that simply swamped local resources. Mallorca had a fine tradition of building skills from architects to stonemasons, but the rate at which not only hotels but also whole settlements and their infrastructures had to be built meant that the only solution was imported labour and skills. As with all such migrations of labour, there had to be 'push' factors as well as 'pull' ones (Salvà Tomàs, 2002a,b). In the Peninsula, rural unemployment and poverty were very high at this time, especially in Andalucia and the Mediterranean coastal lands, and it was from there that substantial numbers of workers were drawn. By 1970, 32% of Mallorca's population had been born in Andalucia, 12% in Catalunya, 7% in Valencia and 6% in Murcia but even distant provinces such as the Canary Islands and Galicia supplied some immigrants such was the attraction of the booming trade. Between 1956 and 1960, migratory balance accounted for just over 7% of population growth in Mallorca, but by 1971–1975 this figure had risen to more than 27%. In the early 1960s, such inward movement was concentrated in the Palma municipality, but by the mid-1970s, nearly 90% was in the *part forana*, that is, in the new seaside settlements. By 1970, 20% of Palma's population had been born in the Balearic Islands but outside its municipality; nearly 28% came from other parts of Spain (Carbonero & Salvà, 1989: various tables). From 7600 employed in hotels in the Balearics in 1964, the number rose to 38,000 in 1973 (Monserrat i Moll, 1990: 99). In construction, employment increased from 9700 in 1957 to 20,600 in 1964, figures that can probably be doubled because of the 'black' economy operating at that time (Buades, 2004: 168). A complicating factor in this search for labour by Mallorca's tourism industry was competition from elsewhere. On the one hand, many other regions of Spain were also beginning to industrialise as well as develop their own tourist activities, and on the other this was also the era of considerable overseas migration of workers from Spain to the growing industrial economies of north and west Europe. Mallorca had to either offer better wages or better working and social conditions to attract these migrants; language and culture were almost certainly a factor.

The impact of this inward migration fuelled by tourism was profound economically and socially. Without it hotels and resorts and new urban centres, the suburban expansion of Palma would not have been possible. While the economic benefits at first sight might seem self-evident, the social effects became apparent only later; by the 1980s, a quarter of Mallorca's population had been born outside the islands. Some of its effects included the impact on language and culture; immigrants rarely spoke Catalan, even fewer the Mallorcan dialect and few appreciated the island's history and traditions. By 1989, some 30% of Mallorca's population older than six years had no measure of Catalan. Island society was moving towards bilingualism under the impact of tourism, and by the 21st century, multiculturalism.

Tourism and Urbanisation

At the heart of the transformation of the island's economy, landscape and society by tourism in this period was the process of urbanisation. A rural, farming community with limited industrial development and with its one primary city was to be completely altered by tourism. In order to accommodate the scale of the growth and its rapidity, massive movements in capital, construction and above all labour had to take place. Buades points to the nuclei of 30 functioning settlements being laid down between 1958 and 1963 affecting 30% of the coastline (Buades, 2004: 163).

The sort of land-use changes described earlier in relation to Platja de Palma were only part of a complex process of urban development. Models and theories based on industrialisation so valuable for understanding 19th-century change elsewhere have not proved sufficient to understand the urban settlements of Mallorca. There are few studies of tourism-led urbanisation (Gladstone, 1998; Mullins, 1991). This was not simply a matter of a set of spatially independent hotels linked to the usual environmental resources of beach and coastline. However, it could be argued that the founding of a large hotel complex on a new beach-front site might act as the pre-urban nucleus for the settlement that was to follow (Carter, 1995: 365). The scale of the shift was – had to be – so large, and in the event so rapid – that it could be achieved only through the creation of a whole new urban system and the expansion of the primate city. Salvà and others (Quintana Peñuela, 1979; Rullan Salamanca, 2002; Salvà Tomàs, 1990: 63–70) have described the growth of this new urban system in graphic terms. Much of the emphasis in their analyses has been on urban population increases, land-use changes and the planning issues that arose from urban growth. Rullan has been especially concerned to address the debate on the forms that urbanisation produced (Rullan, 2002: 352–360), but a more detailed explanation of the processes remains elusive. Just how does a tertiary activity such as tourism bring about the size of urban movement that Mallorca has witnessed since the early 1950s? To what extent can a hotel or a set of holiday apartments with a strong seasonal occupational bias really be a substitute for factories, offices and retailing in the more conventional historical models of urban growth? Are the urban forms and functions produced by tourism truly urban in the usual sense? Clearly, simple supply and demand can sketch in the basic principles and it is possible to describe the physical responses in terms of land acquisition, construction, job creation, population movements, etc but is the process of urbanisation merely the sum of these events or something more? Quintana Peñuela (1979) points to the impetus given by the primate city, Palma, itself a conventional historical city, and the spread of tourism-based urbanisations from there to the coastlands to the immediate east and west, followed by the atoll-like form of a ring of settlements around all the coasts with accessible beaches later on in this first 'boom'. Was tourism alone the engine for urban growth or was the island really part of the general and secular shift in Spanish society towards urban forms, functions and living going on at this time? Rullan's examination of urbanised Mallorca today shows a considerable growth of urban settlements well away from the coast especially along the Raïguer (the irrigated zone between the mountains and the plain) and so it might be possible to speak of an island that is largely urbanised, functionally linked in an almost megalopolitan way with an even more dominant Greater Palma (Binimelis, 1998: 185–203). Certainly, Mallorca cannot be separated from contemporary coastal Mediterranean urban growth in Spain. Some have seen it as part of the broader *arco mediterráneo occidental* (d'Entrement, 1993) and linked to the Californiasation of the Golfe du Lyons (Buswell, 1996: 337).

Urbanisation covers a multitude of environmental elements upon which tourism in this early 'boom' period was to impact. Firstly, as we have shown, there was the land itself: the conversion of agricultural land to urban uses. However, caution has to be exercised here before adopting a condemnatory attitude. The new settlements had by definition to be coastally located and beach-fronted. Historically and economically, such geographical zones had been avoided by Mallorcans and to them had a low value. Where such types of land formed part of historic estates, they were often passed down to younger siblings – often women – who were only too willing to dispose of them to speculators while other relatives concentrated on the agricultural potential of more inland areas. In other cases, the late 19th century subdivision of landed estates into small-scale farms took place, but these, frankly, were too small to be economic and their new owners were equally ready to dispose of them to hotel builders and resort developers. Both of these processes were already underway earlier in the 20th century, as we saw in Chapter 3. In the 1960s, the main difference was the scale and speed of land disposal. Both parties – large landowners and smallholders – greatly underestimated the value of their beach-front properties. Nonetheless, the techniques used by bankers and property developers were often unscrupulous, heavy handed and without concern for the environmental consequences. Buades has pointed to the strong political links between coastal developers, Francoism and local government – 'capitostos feixites i d'aventurers financers' (fascist capitalists and financial adventurers) (Buades, 2004: 164). Some have seen these developments as rewards for support for the regime.

The key investment at the heart of any development was hotel building. In this initial 'boom' period, it was assumed that practically all accommodation for tourists would be in hotels, continuing a tradition built up before the Civil War. The idea of self-catering or apartment accommodation was to come later. Mallorca did not turn its back entirely on camping, another possibility – a holiday form that was proving popular in France and the Peninsula at this time – but the authorities did not encourage it. Fred Pontin's holiday camps were another form.

If the 'mass' in mass tourism can be seen as part of a Fordist model of production and consumption (Bramwell, 2004: 7; Urry, 2002), then it is not solely defined by numbers travelling; it also included the economies of scale that could be derived from hotel accommodation and the manufacture of a standardised product. Some countries relied for many years on the small, family hotel, notably in Britain and France, and in the beginning Mallorca followed a similar pattern. Even when much larger hotels were built, they continued to be controlled by local, family interests. From the small-scale beginnings to the larger units perhaps reflected ideas of gigantism often associated with Fascist architecture as well as current economic ideas. Mallorca was at great pains in this early phase to present itself as 'modern' rather as it had in the first decade of the century. Seguí Aznar suggests that a new style began to emerge for mass tourism that moved away from the Mallorcan style favoured by architects in the 1940s (Seguí Aznar, 2001: 116). By the mid-1950s, hotels accommodating some hundreds of holidaymakers had been built in the 'glass and concrete' modern style. To begin with, the emphasis was on the high end of the market, with hotels such as Hotel Nixe in Calamajor (1957), the Bahia Palace on the Paseo Maritimo in Palma (1955) and the Acapulco in Platja de Palma (1957) set in their own landscaped grounds. The lower-grade three-star hotels usually associated with mass tourism really began being built after 1960 (Photos 4.2 and 4.3). These might be grouped into two principal types: the large-scale hotel situated on the 'front line', that is, on the beach front and in the streets at right angles to the seashore, and secondly, the smaller, low-rise bungalow and chalet-type accommodation surrounding centrally located services such as dining rooms and bars, established in *ciudades de vacaciones*. Examples of the first type include Hotels Fenix and Victoria in Palma (1957 and 1963) and the Hotel de Mar in Cas Català in Calvià (1966). The latter group were especially significant in spreading urbanisation into small, often isolated bays with protected sandy beaches – amongst the most beautiful locations on the island at the time – usually with poor-quality architecture supposedly reflecting Ibizan or Mallorcan traditional forms. Examples include Cala Romantica (in Cala Estay d'En Mas), Mini Follies in Cala Llamp and El Pueblo, Bonaire.

The financing of many of these hotel developments relied on capital from overseas investors as well as Spanish ones, working in partnership with local and Spanish government authorities. British investors were able, by the 1960s, to raise private capital more easily for tourism-related schemes. The sources of Spanish or Mallorcan capital remain somewhat obscure, but the major banks clearly played a significant part. From 1942, the Spanish government had been able to offer bank credit for hotel building and supporting activities (Newton, 1996: 142). This was via Crédito Hotelero referred to in the previous chapter, which provided loans at about 4% or 5% for up to 60% of the total building cost of any hotel for up to 15 years (Buades, 2004: 159). The loan could then be extended by commercial banks at favourable rates of interest. The modernisation of hotels, their furnishing and fitting out and the construction of bars and cafes also benefited from government financial support. Buades has calculated that Mallorca received about 13% of all Spanish government credit support under this scheme, amounting to more than 125 million pesetas financing more than 3000 schemes and eventually providing more than 6000 hotel and other bed places (Buades, 2004:160). To complement this support for private investment, the government also invested heavily in the local infrastructure for tourism – roads, water supply, energy and airports, but this was not widely available until late 1960s (Bray & Raitz, 2001: 68). Pontin's

Photo 4.2 1970s high-rise hotels in Magaluf

Photo 4.3 1970s front-line hotel architecture in Platje de Palma

Pontinental venture in Mallorca was a private one in conjunction with Spanish partners whose

> tastes proved extravagant. I was accustomed to watching every penny ... but exercising the same elements of control on overseas expenditure was far more difficult. There were also exchange control problems and the necessity to obtain permission from the Bank of England for all foreign currency transactions ... but the land was cheap, building costs, providing they were carefully controlled, were comparatively low and catering costs very competitive. (Pontin, 1991: 82)

In the 1970s, Pontin added a hotel to the camp complex and further expanded his activities in Mallorca by acquiring the Belgian S.A. Hotel Club, which also operated holiday villages there (Pontin, 1991: 96).

In these early days, the hotel builders and developers were able to control supply to their advantage:

> ... it was a seller's market par excellence. Armed with government loans from the Credito, top-up finance from the private sector, and interest free loans from the tour operators, they entered into a blizzard of hotel construction the like of which had never been seen before. (Bray & Raitz, 2001: 68)

The greater environmental effects, however, were to be seen once development began to take place largely because of the failure to enforce the admittedly weak legislation embodied in the Land Use Planning Law of 1956. Although planning will be given more attention in Chapter 7, it is worth noting here that in 1963 only eight urbanisation projects were approved under this scheme but Buades calculates that at least 30 were underway by that date (Buades, 2004: 161, 163). Amongst the schemes begun in this era were Son Serra de Marina (Can Picafort) in 1953, Cala Murada in 1956, Cala Blava in 1960 and 'Sometimes' (S'Arenal) in 1963. Also in that year was proposed the 'city of the lakes' in Alcúdia in the north of Mallorca close to 6 km of beach. Such large-scale construction projects required massive amounts of building materials, especially cement, stone and steel, some of which it could procure from Mallorca's own resources by the rapid quarrying and mining of the landscape and building new factories but much had to be imported including from Fascism's sworn enemies in communist east Europe (Buades, 2004: 166).

Negative Impacts

Just as local intellectuals became concerned for language and tradition even at the very beginning of this first 'boom' so too did many soon appreciate the other early negative effects of tourism. Ripóll Martinez identifies four: land speculation, the absolute lack of planning, the destruction of the historic and cultural landscape and the general degradation of the natural environment. Clearly, the increase in tourist numbers from 360,000 in 1960 to 1.9 million in 1970 accompanied by the attendant expansion of the built environment had to be accomplished at the expense of the 'traditional' environment in this decade; there was little concern for 'costs' with, instead, an emphasis on what were perceived to be 'benefits'. Most of the latter were initially to accrue to developers rather than workers and the former were borne by the environment.

Warnings of the environmental effects of this rapid development of tourism in the 1960s came early but were not seriously addressed until the 1980s and after. Although legislation to try to control changing land uses had existed since the 1950s, most of it emanated from Madrid and was largely ignored in peripheral area such as the Balearic Islands. The Lei de Sol of 1956 was well meaning in its intentions, but in practice the sheer scale of the developments of the 1960s led to it being ignored. Rullan has noted that only infrastructural projects such as roads and the airport were really in any sense 'planned'. The doubling of the land area classified as urban between the mid-1950s and 1973 was largely uncontrolled. The municipalities where it took place had been given a greater say in planning decisions from 1963 but had little experience of such pressures and were susceptible to developers' avaricious demands

(Rullan Salamanca, 2007a,b: 22–23). While Mallorcan authorities were not unaware of the growing political significance of the environmental movement, it remained a pressure group activity and was not embedded into the planning system until the 1990s. However, the burgeoning tourist trade in Mallorca demanded improvements in water supply and electricity generation, and their construction had a marked environmental impact. The Gorg Blau reservoir was completed in 1969 and Cuber in 1972. Large as some of these schemes were in this period, their environmental footprint was small in comparison to that of the sprawl of the tourist urbanisations.

A Few Facts and Figures: Temporal and Spatial Patterns of Tourist Numbers, 1960–1973

We have emphasised the rapid build-up of tourist numbers in Mallorca in the early period so that it could rightly be called the first 'boom'. But where did they come from, where did they stay and for how long? How did these statistics change over time?

Flights (one way) to Mallorca increased fourfold from 10,157 in 1961 to 40,603 in 1973, with corresponding visitor numbers increasing nearly nine times from 409,735 to 3,548,358 (one way), illustrating the rising capacity of aircraft referred to above. Table 4.2 shows the proportion of

Table 4.2 Percentage distribution of tourists by country of origin, 1960–1973

	1960	1970	1973
British	23.3	33.3	37.9
German	9.3	19.7	20.3
Spanish	21.6	10.7	10.4
French	13.4	7.5	6.8
Swedish	6.5	4.7	3.8
Belgian	3.6	3.9	3.6
North American	5.0	4.4	2.4
Swiss	4.3	2.4	2.2
Italian	1.8	1.0	1.2
Irish	0.6	0.8	0.7
Various	9.6	11.6	10.7
Total	100	100	100

Source: Conselleria de Turisme (own elaboration)

British and German visitors rising through this period and nearly all other markets stable or declining, producing a pattern of supply that has dominated Mallorca ever since. The dominance of Spanish tourists soon gave way to three quarters of a million British and nearly half a million Germans; the French who had been the second largest group in 1960 with 79,000 visitors were now relegated to about 170,000. Tourist beds rose from 22,000 in 1960 to 266,000 while in terms of bed-nights the total rose from 4.75 million to more than 50 million. The length of stay remained fairly constant at between 12 and 13 days on average, suggesting that the dominant mode was the fortnight package holiday. The increasing numbers of tourists also consolidated even further the seasonality effect in July and August. Although there were national variations in this concentration, the dominance of traditional manufacturing industries in the economies of Germany and Britain in this period led to traditional factory (and school) vacation patterns.

Chapter 5

From Crisis to Crisis (1973–2010) with a Continuing Boom in Between!

The First Crisis

If the 1960s had been the era that had witnessed the most rapid rise in tourist numbers in Mallorca, then the euphoria that accompanied it was soon to be brought to a sudden halt by the rapid stagnation of the fortunes of the European economy in the early 1970s (Figure 5.1). In autumn 1973, the Egyptian and Syrian armies attacked Israel on Yom Kippur, October 6th, beginning the fourth Middle East conflict between these nations. The Soviet Union assisted both Arab countries materially, and the United States aided Israel. The war ended in something of a stalemate but with Israeli gains in the Golan Heights to the north and the encirclement of the Egyptian third Army to the south. On 22 October, the UN Security Council succeeded in imposing a ceasefire.

The War itself might be seen as a geographically localised affair, but its effects were to be global. In retaliation to the American support of Israel during the war, the Arab members of OPEC, including the oil-rich state of Saudi Arabia, decided to reduce oil production by 5% per month. The Americans continued to support Israel with a major allocation of arms supplies and more than $2 billion of assistance. Saudi Arabia and other oil exporters declared an embargo against the United States and other Western industrial states, leading to an energy crisis characterised by a price shock. The rapid and near quintupling of world oil prices triggered a stock market collapse in the ensuing months and, indeed, changed relations between the West and the Arab world for decades to come.

For the West European economies, this combination of oil price increases and a fall in stock market values led to a tightening of economic belts. One of the first casualties was a marked reduction in disposable incomes for things such as holidays, accompanied by a degree of fear in the dangers of international travel. Transport costs for charter airlines rose alarmingly; the days of the cheap flight were not to return for another two decades. For Mallorca and the Balearic Islands, it meant a reduction in demand for package holidays for the 1974 season. The year ending in the summer of 1973 had seen more than 4.3 million tourists coming to the Balearics; in the same period a year later, this number fell to 3.9 million. For Mallorca itself, the figures were, respectively, 3.5 million

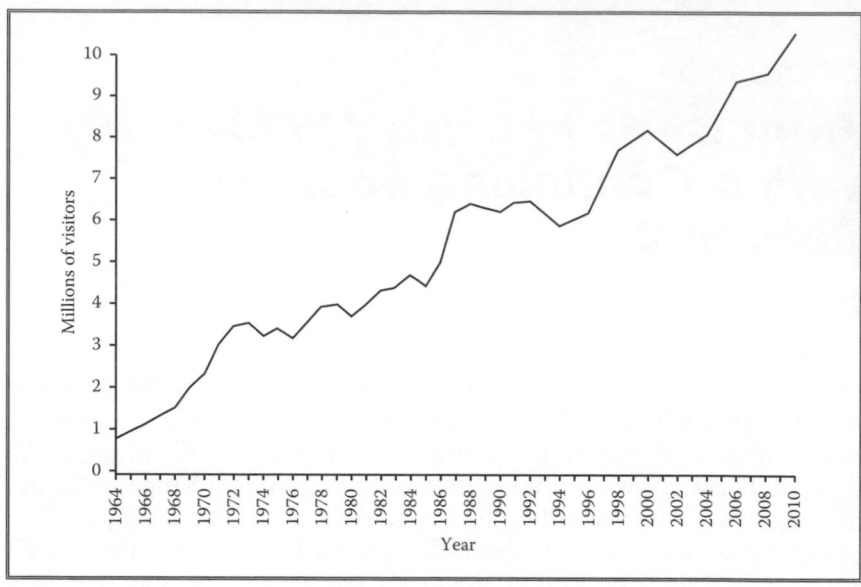

Figure 5.1 Graph of tourist numbers 1970–1996
Source: Dades Informatives

and 3.2 million. This was the first time such numbers had fallen from one year to the next since the 1950s (Seguí Llínas, 2006: 11). Not until the end of this turbulent decade did numbers coming to Mallorca by air rise to 4 million.

Britain at that time was the major source of tourists to Mallorca, and it was essential for the island for that supply to be maintained under these difficult conditions. Unfortunately, a combination of factors led to the demise of some of the country's principal tour operators, notably Court Line, which provided air services for one of the largest holiday firms, Clarksons (Bray & Raitz, 2001: chap. 8). With the collapse of Court Line in August 1974, 49,000 of its clients were left stranded overseas, 40,000 of them on a Clarkson package, many of them in Mallorca (*Flight International*, 22 August 1974: 198). This was the largest such implosion in British aviation history. In an attempt to try to safeguard against further losses, the holding company Court Line Group first bought Horizon Holidays and then Clarksons, neither of whom could now survive. A contributory factor was the tour operators' attempts to provide the lowest possible price for package holidays to Mallorca, often below cost. Subsequent investigation by the British Board of Trade concluded that such firms were badly managed financially with assets too widespread. One of the unintended outcomes of this collapse was the introduction of the ABTA scheme to fund the return of passengers stranded abroad by such business failures. The impact of these events was catastrophic in the short run, but remarkably demand from Britain soon recovered and other airlines and tour operators stepped in to fill the vacuum created. Nonetheless, Mallorcan hoteliers did begin to question the robustness of the British market and increasingly turned to West Germany although it was not until 1990 that German visitors exceeded British ones.

Franco's death in 1975, which led to a phase of political instability and rising inflation throughout the 1970s, seemed to make tourism a less dependable engine for continuing economic growth in Spain. In Mallorca, the combined effects of the first half of the 1970s may have seemed dramatic, but, in fact, the island began to recover its tourism crown rather more rapidly and successfully than many other parts of Spain. This was almost entirely due to the reactive entrepreneurialism of the Mallorcan hoteliers and local politicians who quickly mounted marketing expeditions to Britain and Germany to advertise the island's attractions, to reduce package tour costs where possible and to try to create an atmosphere of 'business as usual'. However, the brief decline of the mid-1970s did act as a stimulus for change in the nature of the tourist experience. It also brought home to the Mallorcans themselves that too much of the industry was now controlled in effect by overseas capital in the form of the tour operators, the important intermediaries between supply and demand. To some Mallorcan observers, the 1970s saw this as confirmation of a long-held belief that tourism had become a kind of colonialism (Arnau i Segarra, 1999: 40; López Bravo, 2003). What followed in the 1980s was in many ways the beginning of new dimensions to Mallorcan tourism.

The industry began to recognise that a number of its characteristics in Mallorca could no longer be relied upon for an ever-increasing mass tourism market. There was a suspicion that the latter stages of the Butler model were being approached if not actually reached. Slowly, analysis began to reveal a number of characteristics that needed remedying. Firstly, the narrow seasonality needed to be broadened out to include more visitors in early summer and autumn, the so-called shoulders of the annual distribution curve. Secondly, the almost total reliance upon the three-star package tourist hotel with rather basic services would no longer satisfy a more discerning market. Different seasons and a different clientèle would demand a different kind of provision. Thirdly, the infrastructure supporting tourism was now neither efficient nor sufficient. Fourthly, the environmental and social impacts that had largely been ignored in the race for growth in the 1960s were now higher on the agendas of both tourists and locals alike. Water supply, electricity generation, transport and public spaces had all come under considerable pressure in the previous two decades. Fifthly, the growth of the island's population from 363,199 in 1960 to 460,030 in 1970 and on to 551,215 in

1981 was beginning to put additional pressure on resources irrespective of tourism's demands. While tourism may have stimulated much of this growth, it became clear that it could not sustain new levels of employment by this means alone. The increase was brought about principally by continuing immigration, leading to a much more diversified demographic structure in which a new generation of Mallorcans not born in the island and without the characteristics of Mallorcan culture – especially linguistic ones – was beginning to uncover new political tensions. Palma became increasingly dominant in the urban structure with more than 54% of the island's people by 1981, which had many ramifications for economic and social activity (Salvà Tomàs, 1986: 7, 9, 11). These changes were not immediately recognised by either tour operators or hoteliers. Politicians and economists could no longer hide behind the assumption that tourism was somehow exogenous to the life of the island, an add-on to historical evolution; by 1980, it was all-pervasive. As we shall see later, Mallorca was one of the first regions in the Mediterranean that could describe itself as quintessentially the creature of the tourism industry. Many of these weaknesses identified by the early 1980s were to persist right through to the 21st century, leading to serious questions about the applicability of the Butler model to Mallorca (Coles, 2006: 55; Picornell & Picornell, 2002a: 33). Growth in numbers began to seem unstoppable.

Within the structure of Mallorca's tourist accommodation, the post-1973 era signalled a halting and tentative move away from the three-star hotels that had dominated holidays in the 1960s. A major shift in the supply side of the equation was the fall in hotel building, relative to other forms of real-estate investment. In the short term, this meant a loss of employment opportunities particularly amongst mainland immigrants, and a stabilisation in the work for hotel staff when the island gradually began to move some of its accommodation emphasis to self-catering apartments. The total stock of bed spaces increased by 26% (1973–1981), but most of this (86%) could be accounted for by additional places in apartments. This may have lowered costs for visitors but again, it reduced local job opportunities (Alenyar, 1990: 30). More important, it signalled a move away from the hotels and introduced a new element into the townscape of the resorts. This competition within the Mallorcan market naturally brought a response from the hotel owners. In the 1960s and 1970s, hotels were built and run by a multitude of small and independent operators who benefited from credit from local bankers. By the late 1970s and early 1980s, the capital locked up in these investments was largely exhausted and many hotels needed refurbishing to maintain standards (Photo 5.1). Small- and medium-sized enterprises began to find it difficult to raise further capital at prices that they could afford. To begin with, the solution lay in cooperation rather than competition between such hotels with the establishment of COFEBRA (Cooperativa

Photo 5.1 Example of publicly funded pedestrianisation in Cala Millor

de la Federació Balear) in 1977 (Mata Pastor, 2002: 92). This had limited
success and a process of merger, acquisition and takeover gradually built
what might be described as larger 'chains' of hotel companies. In
addition, it now became more profitable to develop higher-grade four-
and five-star hotels to accommodate a slowly growing, more discerning
market. In Mallorca, in 1987 there were still 198 one-star hotels, 147 two-
star, 191 three-star, 49 four-star and only six five-star hotels. By 2008, the
respective figures were 33, 60, 293,146 and 23, showing the considerable
decline in poor-quality accommodation and a marked increase at the top
end of the market. Measured by bed-nights, the ubiquitous three-star
hotels *increased* their share between these two dates from 25.7% to 30.5%
but the four- and five-star hotels together doubled their share from a
meagre 8% to 17.4%; by 2008, the one- and two-star hotels had virtually
disappeared. In 1987, just under two thirds (62.8%) of Mallorca's
accommodation capacity was in hotels; 20 years later, this had fallen to
a little over half (52.5%). The proportion in older-style apartments fell
too, from 19.6% to 16.2% (Table 5.1). In spatial terms, the patterns
established by the late 1970s with concentrations of all accommodation in
the coastal zones of Palma/Llucmajor, Calvià, Santa Margalida/Muro,
Pollença/Alcúdia and Sant Llorenç/Son Servera were not to change
substantially.

Table 5.1 Mallorca hotel capacity by type, 1997 and 2008

Accommodation type	1997			2008		
	Establishments	*Units*	*Places*	*Establishments*	*Units*	*Places*
Apartment types 1–4	432	20,445	55,733	387	17,140	46,340
Three-star hotel	279	47,335	89,737	293	46,307	86,965
Four-star hotel	65	12,189	23,099	146	23,398	44,352
Five-star hotel	5	573	1,129	23	2,651	5,312
Total	349	60,097	113,965	462	72,356	136,629
Three-star apartment hotels	74	14,520	33,593	85	16,137	37,241
Four-star apartment hotels	14	3,239	7,136	74	14,894	32,504
Five-star apartment hotels	0	0	0	3	226	448
Total	88	17,759	40,729	162	31,257	70,193
Agrotourism	52	259	523	151	998	2,082

Source: Dades Informatives (1997, 2008)

Sastre provides some useful data for the Balearic Islands as a whole to demonstrate the changing structure of hotels during this period. While the largest hotel groups increased their share of the market, they did so by expanding/acquiring more middle-sized hotels especially those with 300–400 beds as Table 5.2 shows.

Accommodation in apartment hotels did not begin to be officially recorded until 1994, but its capacity then doubled by 2008 to about one quarter of the island's bed spaces in the period 1994–2008 when overall capacity increased by only just over 10%. This 'third boom' era saw a marked cultural and structural shift in what holidaymakers wanted from their accommodation (or more accurately perhaps, what they were offered) but, importantly, the three-star hotels catering to the package tourist succeeded in remaining the backbone of Mallorca's holiday trade. Although the rate of their construction slowed down from the mid-1990s, their number increased by more than 100 in the period 1987–2008 and their average capacity has remained remarkably constant at about 320 bed spaces each.

So, the proportion of the total accommodation capacity of Mallorca in all types of hotels as opposed to other accommodations has fallen in the last two decades. At the same time, the inflationary pressures of the continuing boom in tourism inevitably pushed up wages and salaries and so moves to reduce costs of labour relative to other costs became a pressing issue. This was partly the result of the liberalisation of labour laws post-Franco and the growing strength of the trade union movement. In the wake of the 1973 price shock, there had been a growing number of strikes, aggravated in part by the increasing labour demands of tourism, which seduced more and more workers from agriculture and fishing and to a lesser extent from a whole range of secondary industries.

Table 5.2 Hotel size changes from 1970s to 1990s

Hotel size by no. of places	All Balears (%)		
	1968	*1980*	*1992*
100–200	27.4	19.2	14.1
300–400	14.3	14.3	16.2
No. of places in a hotel group	Total spaces in Balears (%)		
	1975	*1986*	*1992*
2–3000	7.0	1.2	9.2
7000+	5.6	12.7	13.6

Source: Modified from Sastre (1995: 98)

As the tourism industry continued to grow through the 1980s, local producers could not meet the demand and so more and more goods and services had to be imported. In 1980, the 12% of the workforce engaged in the primary sector had fallen to just 5% by 1990; secondary sector employment fell from 27% to 26%. On the other hand, the continuing seasonality of the industry did make employment in more 'stable' sectors such as local government more attractive. Together with the rise of food retailing outlets, especially supermarkets such as Caprabo (later Eroski), Lidl (a German company) and Spar, these three factors combined have encouraged the self-catering and residential sectors of tourism in Mallorca during the last 30 years.

By the early 1980s, the appearance of many of the urbanisations that had grown up on the coast was becoming rather tired and run down. The original private investment that created them had naturally focussed on the hotels themselves and the small sites they occupied. In some resorts, especially those whose origins went back to the earlier town planning movements in the island, there was more emphasis on roads, public open spaces and general infrastructure from the beginning, but in many of the 'boom' resorts on the south coast such things were neglected, a case of the economist John Kenneth Galbraith's 'private wealth and public squalor'. It was assumed, wrongly, that the municipalities would pay for those parts of the tourist environment that were thought to be in the public realm but many, such as Calvià and Palma and Llucmajor, initially saw the income generated by tourism as a means of funding other priorities often located in the old interior centres of the *municipios* and, of course, as a means of lining private pockets, for corruption was said to be rife at this time. The competition between local authorities for a share of tourism wealth during the two decades to 1980 had little regard for the longer-term benefits. By the mid-1980s, the returns on capital in the private and public sectors were nearing exhaustion (Cals, 1990: 129). This kind of local neglect of the externalities of tourism had taken its lead from the national political situation under Franco and during the economic chaos of the transition to full democracy from 1975 to 1981; it was not until the success of the autonomia movement in 1983 that regional and island government was able to address some of these issues. It was only in this century that the public sector recognised the necessity of investing in the visible infrastructure to improve the townscapes of resorts. Perhaps the best example has been the construction of the new promenade in Platja de Palma and an ealier example from Cala Millor (Photo 5.2). Price controls were lifted in the mid-1980s, and the political management of tourism transferred from the Ministry of Information and Tourism in Madrid to the Govern Balear in Palma. At the same time, the powers of the island council – the Consell – similarly increased. One of the early results of devolution was the first *Llibre Blanc de Turisme a les*

Photo 5.2 Recently refurbished high-rise hotel in Cala Millor

Illes Balears published in 1987, a joint study between the Govern Balear and the University of the Balearic Islands. It was the first really comprehensive study of the tourism industry in the islands. Mallorca, like all of Spain, probably began to experience the effects of too many layers of political representation. Until the early part of this century, it is true to say that the lowest layer – the municipality – had too much discretion in planning terms theoretically but usually lacked the power to resist the influence that could be brought to bear by developers, despite the emergence of much stronger island-wide planning controls and legislation (Pack, 2006a: 124). For example, in the case of Calvià municipality '... the problems of badly co-ordinated development, unlimited construction, growth and the unsustainable use of natural resources were clearly evident'. In 1960, Calvià had 3000 residents, 6800 bedspaces and 112 tourist establishments. By 1997, there were 35,000 registered residents, 50,000 de facto residents, 58,000 bedspaces and 256 tourist establishments (Dodds, 2007: 300).

The first Llibre Blanc (1987) showed, for example, that the move towards apartment building and away from hotels was often at the expense of proper planning control. By 1986, of the 257,210 bed spaces in Mallorca, 65% were still in hotels but 32% were now in apartments, of which more than half were not officially approved. This was especially true in the 'newer' resort areas of the east coast, the south-west and the northern bays. Platja de Palma and Palma itself, historically dominated by hotels, had managed to keep out the worst kind of illegal apartment building (Table 5.3) (Buswell, 1996: Table 12.6, 324), which, of course, does not exclude unapproved hotel building.

In addition to the restructuring of the accommodation sector of the industry, the organisations responsible for the majority of vacations in the Balearic Islands, the tour operators, also underwent considerable change. Most notable was the process of vertical integration, which led to a small number of large companies owning and operating the whole process of holidaymaking for many consumers. From booking to delivery the tourist was in the hands of one organisation. Such companies in the late 1980s and much of the 1990s had their own High Street travel agents, travel-to-airport buses, airlines, hotels, travel-to-resort buses and even their own side-tour operations. Tourists became truly packaged (Sastre & Benito, 2001: 71). By 1998, 92% of arrivals in the Balearic Islands came via tour operators, only 8% being independent. Such tour operators included the familiar names of Thomson, Airtours and First Choice from the United Kingdom and TUI, NUR-Neckermann and LTU from Germany. During the last decade, many of these have merged further or ceased operation. The considerable increases in German numbers coming to Mallorca might be explained by the takeover of British tour operators by German firms, leading to circular and cumulative growth in the number of German visitors. These larger tour operators helped fund hotel construction in 1970s and 1980s, and further vertical integration resulted in the joint ownership or takeover of more hotels, mostly in the three-star category, in 1990s (Sastre & Benito, 2001: 78). This process of rationalisation was by no means unique to Mallorca; it was part of a worldwide shift as the tourism industry became globalised.

Second Homes

In Chapter 8 we will examine the ways in which the tourist market in Mallorca has been diversified to include alternatives to the sea, sand and sun holidays concentrated in the summer months. One aspect of this process that deserves attention here, however, is the growth of second homes and residential tourism. What appears to be an apparent contradiction in terms actually provided Mallorca with a new sector or dimension to its economy and a new source of social tension. While at the macroeconomic level the Balearic Islands remained one of the richest parts of Spain, their income from tourism fell in real terms between 1987 and 1992. In 1989, gross domestic product per inhabitant had been 35.65% above the Spanish average, rising to more than 42% above the Spanish average in 1993 (Seguí Llínas, 2006: 17). Some of these surpluses

Table 5.3 Mallorca location quotients based on 2008 data*

	Palma	S'Arenal	Costa de Llevant	Alcúdia Bay	Pollença Bay	Costa de Tramuntana	Costa de Ponent	Remainder
German	1.16	0.99	0.76	1.33	1.04	0.83	1.44	1.24
British	1.44	1.55	1.27	0.92	0.84	0.96	0.66	0.98
Spanish	0.59	0.65	2.5	1.65	1.33	2.43	1.31	0.81
All foreigners	1.17	1.17	0.88	0.92	0.95	0.89	0.95	1.05

*1.00 = normal distribution; < 1.0 = fewer than that area should have, e.g. A'cudia has more Germans than might be expected, but fewer British tourists.

Source: Dades Informatives (2008) and own calculations

were expended on second homes in a set of spatial process that was found in other parts of Spain (Barke, 2007; Casado-Diaz, 2004).

During the early years of tourism development in the 1960s and 1970s, most of the second homes were acquired or built for local residents who wished to escape inland from the disbenefits of coastal development or from living in overcrowded Palma, assisted, of course, by a growing permanent population putting more pressure on the traditional housing supply. Often such second homes were renovated farm and village houses once abandoned by the urbanising population of the 1960s. The more affluent urbanites had also colonised parts of the accessible coastline for their weekend retreats. They were the province of the better-off who sought out areas of high landscape value such as in Tramuntana and overlooking areas of unspoiled coastline. By the late 1960s, the demand was coming from the new residential or partially residential tourists whose location of choice was often conditioned by access to the facilities of Ciutat and so the outer areas of the municipality of Palma had the highest density of second homes at this time. The next decade – the 1970s – witnessed a doubling of the number of second homes to 57,000. For non-residents, two similar variables remained important: to be near the sea and to have good access to Palma. Areas with lower intensity of coastal tourism, such as Campos and Felanitx, became attractive, but the major concentrations were in Calvià and Alcúdia. In the interior, the most important locations were within a half-hour journey of Palma and so access to main roads was important. Municipalities such as Marratxí and Santa Eugènia benefited from this, especially on non-irrigated, low-value farmland. By the early 1990s, unofficial estimates suggested that Mallorca had almost 90,000 second homes (Salvà Tomàs & Binimelis, 1993: 74–75). By then, remoteness in areas of high landscape value became most desirable, especially for the growing number of extremely wealthy partially residential tourists such as Michael Douglas's controversial home (S'Estaca) on the former estate of Ludwig Salvator. Many so-called personalities first came as tourists but returned to buy property or move their yachts here (Institut Balear de Turisme, 2001).

For many visitors, and especially those from better-off backgrounds who do not want hotel or apartment accommodation for holidays, there has been a growing demand for villa rentals, now booked via the Internet usually on a weekly or fortnightly basis and accessed by cheap flights and car hire. It is another sector that is poorly regulated and so data are sparse. It has been estimated that in 1975 there were perhaps 4000 of these villas, a number that rose to about 39,000 in 2000 (Blázquez Salom, 2005). In the same period, only about 7000 new licenses were issued for such buildings. From the 1980s, this sector was part of the spatial diversification away from the coast and so the presence of a swimming pool was a vital characteristic. Its development may also have encouraged a social diversity away from the mass of tourists, but often its environmental impact was quite high, helping to convert parts of the attractive country-side to an almost suburban character. Intense users of water (gardens and pools) and electricity (air conditioning), they are largely self-contained and contribute little to the local economies apart from providing a demand for maintenance services. Unlike more permanent residential tourism, the majority of these properties remain empty for long periods in the low season. A large proportion appear to have overseas owners.

Immigration and Tourism

In an earlier chapter, it was noted that by 2008 there were almost as many people in the Balearic Islands born elsewhere as in the islands – 579,000 versus 492,000. Of the latter group, almost half (222,000) were born outside Spain (Diario de Mallorca, 21 June 2008). Salvà has divided these non-Spanish immigrants into two groups: those from Africa and Latin America who are generally less skilled labour migrants, and north Europeans who tend to be residential tourists and retirees (Salvà Tomàs, 2002a,b). This is a rather simplistic division because one group of Europeans that probably belongs with the migrant workers' category is the south-east Europeans, Bulgarians, Albanians, Rumanians etc. who number a remarkable 18,000 plus. In Mallorca, geographically speaking, those born outside the Islands are to be found almost everywhere but the principal concentrations (those with more than 41%) are in the municipalities on the south coast and in the north-west and north-east corners of the island. Only nine inland municipalities have more than 80% of their inhabitants born within the islands (Salvà reported in Diario de Mallorca 14 November, 2009). However, of the north European immigrants (who may or may not have residential status), many will be residential tourists. Their decision to take up residence in the Balearic Islands has been conditioned partly by what Salvà calls the 'lure of the Mediterranean' and its quality of life, which in itself is a product of previous visits as conventional tourists. They are usually also influenced by property company literature and information supplied by friends and relatives. The geographical and social attractions are seen as antidotes to the excesses of urbanisation in their home countries and a rather romantic view of the Mallorcan countryside. They tend to form isolated colonies of social contacts using only supplies of services to do with housing and utilising only rubbish collection and sanitation from local municipal services. The municipality of Palma has proved an attractive location for homes for these so-called permanent residents. Despite Salvà's claim, however, it would probably be best not to describe this movement of Germans, Britishers and others as part of a counter-urbanising process emanating from north European cities since the majority do not reside in Mallorca's smaller towns; they are part of a wider urbanisation of the countryside. It is the North African and Latin American immigrants who have moved into the cheaper accommodation to be found in Mallorca's towns. Jacqueline Waldren's detailed anthropological study of Deià from the 1950s to the early 1990s suggests that over time 'outsiders' can become 'insiders' and that a too hurried categorisation of new residents as merely 'colonists' is too simplistic (Waldren, 1996: 247–249). As she pointed out in 1996, the danger for 'insiders' is that they may soon be outnumbered by 'outsiders' in some localities. Taking the wider view of inward migration to include Peninsular Spanish, the population data quoted above suggest that that day is almost upon the Mallorcans.

Second-home owners confuse the picture because of their sporadic presence in the island, their disregard of local rules (and taxes!) and regulations (often out of ignorance rather than malicious intent) and their linguistic limitations. If there are 'colonists' in Mallorca they are more likely to belong to this group, but without further detailed sociological research at the household level it is difficult to make meaningful judgments. López Bravo's analysis that suggests that 'locals' and incomers get along pretty well is perhaps a little too premature and naïve (López Bravo, 2003).

This long period (1974–2010) has witnessed at least two booms with two noteworthy disturbances in the inexorable upward growth in numbers of tourists – in 1991–1992 and more recently since 2008. Visitor numbers to the Balearic Islands rose from 3.4 million in 1975 to 12.5 million in 2008, with 83% of the latter in Mallorca. From the late 1980s, the number of establishments catering to tourist accommodation has remained fairly stable in Mallorca at around 1600 but the bed-spaces have risen by 23%. The peak of bed-nights of 95 million was reached in 2006. Large numbers of apartments and second homes have been built, many illegally. The population of Mallorca has grown numerically from 492,000 in 1974 to 760,000 in 2009, and its composition has become much more diversified both socially and ethnically. The infrastructure that supports this massive growth has likewise expanded; railway lines abandoned in the 1950s and 1960s have been or are being reopened; a new Metro to the university has been completed; Son Sant Joan Airport has been completely redeveloped and is likely to undergo another expansion shortly. The rapid transformations of economy, society and the environment that began in the 1960s have continued and so dependence on tourism has deepened and intensified in so far as it has led to new and expanded and related tertiary industries. Until the 1990s, urbanisation dominated the spatial patterning, but in the last 20 years there has been a counterurbanising and ruralising of many aspects of Mallorcan life.

The New Tourists

While the emphasis still remains on the conventionally defined package tour, the tourism industry in Mallorca has seen in the last decade a remarkable shift in the ways in which many tourists are arranging their holidays. Two technical and business innovations have moved the organisation of some holidays from the corporate package to the individualised one, a move sometimes described as the 'democratisation of tourism'. The no-frills, low-cost airline and the internet have altered, and are altering, the way many are managing their time abroad. By 1995, Sastre was able to identify the apparent inflexibility in the organisation of Mallorca's tourism market, with most hotel businesses largely unable to break free from the control of the tour operators particularly with regard to price and publicity. Only the larger chains were able to resist tour operator pressures (Sastre, 1995: 123). In the last 15 years, this position has changed somewhat, and in three directions. Firstly, the small number of large hotel chains has increased their oligopolistic position and so their bargaining power with tour operators has increased further. Secondly, the emergence of the low-cost, no-frills airlines has undermined the hold of the tour operators' airlines and national carriers to such an extent that many of them now offer not only seat-only deals but act *ipso facto* as low-cost carriers themselves. Thirdly, the widespread use of the internet and its now massive content, especially by the British tourists, has enabled the High Street travel agent (often part of a tour operator's company) to be bypassed, allowing both flight and hotel booking direct with hotels or through online travel companies. For an example of the assistance available to online bookers from the United Kingdom, see http://www.travelcontentonline.co.uk/booking-your-majorca-family-holidays-online/ (accessed 22 May 2010).

Research has shown that at least in the early stages of this revolution these new business innovations as applied to Mallorca and the Balearic Islands tended to benefit the larger hotel companies because they were more attuned to the idea of change through their more advanced managerial systems and standing; smaller, often family-run hotels were slower to adapt to new practices. A second characteristic of the use of the Internet to book holidays was a noticeable increase in its use seasonally with the low and mid-seasons benefiting most (Garau Vadell & Orfila-Sintes, 2008). CAEB also found that agrotourism establishments benefited too; the use of small, specialist travel agents remained important (CAEB, 2002: 21). Although Table 5.4 does not prove the use of the internet to book a flight/holiday, it does show interesting variations in the use of the internet by different nationalities.

Although the business model for low-cost, no-frills carriers such as easyJet, Ryanair and AirBerlin is different from that of the charter

Table 5.4 Internet and nationality

	Use of internet	*Package holiday (Balearic Islands)*
German	73.7%	54.3%
British	78.3%	46.3%
Spanish	69.8%	37.5%
All tourists to Mallorca	74.5%	49.5%

Source: Dades Informatives (2009) and own calculations

companies that were so important to Mallorca from the mid-1950s onwards, their appearance in Mallorcan airspace from the late 1990s owes much to Courtline, Horizon and Thomson (Britannia Airways) – the early pioneers. Air Berlin inaugurated flights to and from its newly created hub in Mallorca to four locations in Germany in 1999 (Vera Rebollo & Ivars Baidal, 2008). In 2009, the Balearic Islands were the destination for 23% of all low-cost flights to Spain; 84% of passengers to the islands using such flights came from Germany and the United Kingdom (Instituto de Estudios Turísticos, 2009: 32). Today (2010), Son Sant Joan Airport hosts 90 different airlines and all but a handful might be classified as not being low-cost carriers. One hundred and twelve European airports can be reached from Palma, 27 of them in Spain. Between 1999 and 2007 (a peak year before the onset of recession), flights to and from Mallorca rose from 166,997 to 197,384, an increase of 15.4%, largely attributable to low-cost carriers. Of course, not all these flights are solely for tourists; indeed, one of the successes of the low-cost carriers has been their penetration of the business market. The use of air freight has also increased dramatically. As an indication of the decline in winter traffic following the recession, data for January 2010 show that flights were down 6.4%, passenger numbers down 4% and goods by air down 15.7% over the previous period, 2009 (Aena, 2010).

We have shown in this chapter that the package holiday in three-star hotels in mid-summer Mallorca remains dominant, but a third aspect to the 'new' tourists is their desire to participate in activities away from the coast and often in mid or low season. In Chapter 8, we will examine in more detail attempts to diversify the tourism product in order to attract a different ('new') clientèle. Some have seen the future of tourism in Mallorca in this third, composite sector (see Chapter 9).

Chapter 6
Environmental Impact and Sustainability

Introduction: Rising Environmental Consciousness

In our examination of the growth of the tourism industry in Mallorca over the last 100 years, we have noticed an increasing concern for its effect on the environment. In Chapter 2, following traditional lines, the environment was seen as a primary resource for tourism in almost any locality concerned with 'sun, sea and sand'-type of holidays. Two elements contribute to this: climate and coastline, with alternative, often upland, environments offering occasional distraction. However, neither of these two elements is constant. Climate is subject to long-term variability and the associated 'weather' to short-term, often diurnal, variation. In many ways, the coastline is subject to more complex processes, some natural, such as erosion or longshore drift operating over long time periods, and some the product of human interference. On the one hand, this might consist of the impact of the pressure of visitors on natural resources, for example, the degradation of beaches and dunes by the sheer weight of numbers. On the other hand, this might consist of the effects of coastal engineering, such as sand replenishment, groyne construction and the effects of harbour and marina construction, to control physical and human effects. Mallorca has been subject to all of these natural and human processes, but because of the scale of tourism in many parts of a small island their impact has often been disproportionate compared with that in other parts of the Spanish coastline (Oreja Rodriguez *et al.*, 2008).

Consideration of climate change is a much more recent policy issue. The perception from late Victorian times that Mallorca's Mediterranean climate was a 'constant' was rarely questioned. It formed perhaps the principal attraction for a century, the key element in the three 'Ss' of south European holidaymaking. There may have been seasonal variations characteristic of the Mediterranean climate and day-to-day variations in local weather in the summer season, but the 'sunshine' was a constant, especially after the social construction of the importance of a tan for the pale-skinned north Europeans was established in the 1930s. The threat of long-term climatic change in the 21st century and beyond has only recently been recognised, and its possible impact on Mallorca's tourism industry begun to be assessed.

If environment is a resource for tourism, then, as in any economic activity, that resource has to be maintained on the supply side of the equation. If its value declines, for whatever reason, it has to be managed or substitutes found. It is this notion that the environment for tourism must be sustained that has engaged Mallorcan minds more than any other topic in recent years, but it is a topic that needs considerable elaboration away from the simple supply and demand model attached to tourism alone (Butler, 1998: 28). Sustainability is a simplistic notion that readily appeals to politicians, and some planners, for obvious reasons: it suggests that through intervention, management and engineering, the decline of the environmental resource base can be halted or even reversed. But environment is a system-like event in scientific terms. The systems of tourism are inseparable from wider socio-economic systems such as demographic change, urbanisation and non-tourist economic growth. And these systems operate on a much wider geographical scale than Mallorca alone. Indeed, they are especially Europe-wide in their operation and increasingly global. Mediterranean tourism implies movement from one climatic regime to another, and this very movement itself has an environmental effect on the country of origin and en route. As so much of that movement to Mallorca is by air, the current concern for its impact on global warming comes as no surprise. In other words, the tourism industry is simply a subset of wider social, economic and geographic systems. Sustaining some aspects of the touristic environment alone might be possible but other aspects might not. The Mallorcan economy and the island's social life might be dominated by tourism, but as the island has developed over the last 50 years, elements that have grown up on the back of tourism have environmental impacts of their own. The two classic examples in Mallorca are urbanisation and car ownership, both initially dependent upon tourism but now undergoing self-sustaining growth. And at the heart of both of these has been the demographic transformation of Mallorca from an agricultural economy and society that suffered from emigration to a tertiary/quaternary service economy that has developed in a vastly expanded urban milieu, historically fuelled by immigration. The resources for and of tourism remain central and crucial in any sustainability debate, but they now have to be matched by concerns for the environmental impact of a massively increased local population and the non-touristic economy that supports it. What is clear in the recent historical development of tourism is that the industry has shifted from a product-centred activity to a consumption-focused one. Until many of the original investments are exhausted, there will be little new hotel building, for example, but instead, resource refurbishment or conversion. The stock curve of many of the production-side elements of the industry will probably remain fairly constant in quantitative terms as the island nears capacity. Instead, much future investment will have to concentrate on managing the resource base for indigenous, non-touristic activities as well as for the needs of visitors. This emphasis on consumption has quite different meanings for resource use and sustainability.

An example of this kind of effect can be found in the debates that Mallorcans have had about energy supply and the decisions made to link the island with the Peninsula by high-tension undersea electricity cables and a natural gas pipeline. The demand for this increase in energy supply may show seasonal variation because of the tourism industry, but the resident population now makes demands all year round. The choice, in environmental terms, was between expanding Mallorca's own generating capacity with its attendant pollution or locating extra capacity in someone else's backyard (www.siemens.com, 9 October 2007). The Balearic Islands are making efforts to increase energy supply from non-conventional sources, the most important of which is the conversion of urban waste into electricity. In 2007–2008, more than 134,000 KWh of electricity was generated this way. Wind power is not really an option in this Mediterranean environment, unlike photovoltaics whose output is rising rapidly but providing only a small proportion of total electricity output.

Demographics

As consumption is now significantly more important, a principal variable in the resource/sustainability equation is population and so one of the major concerns of environmentalists researching tourism in Mallorca has been focused on the number of people on the island at any one time. However, what Mallorca has had to contend with has been the absolute increase in numbers from whatever source or time of the year. Calculating the total numbers is complex. Clearly, there are two elements to this population: the resident and the transient, the latter, of course, primarily being the tourists but which may include certain recent immigrants and seasonal workers who, theoretically at least, may return to their country of origin at any time. The last 20 years have seen an unprecedented rise in the population and its rate of increase (Figure 6.1). In 1998, there was a registered population in the Balearic Islands of 797,000, a number that has risen to more than 1.1 million today, but note that this and subsequent figures do not include the unknown number of non-registered people living in the Balearics. Natural change (births minus deaths) has contributed a relatively small part of this increase although its rate has accelerated in the last decade. From 1987 to 2006, the natural increase was 45,342. Much more significant has been the increase in net migration (immigration minus emigration). From 1989 to 2006, net migration was 136,616, with more than 100,000 of this number

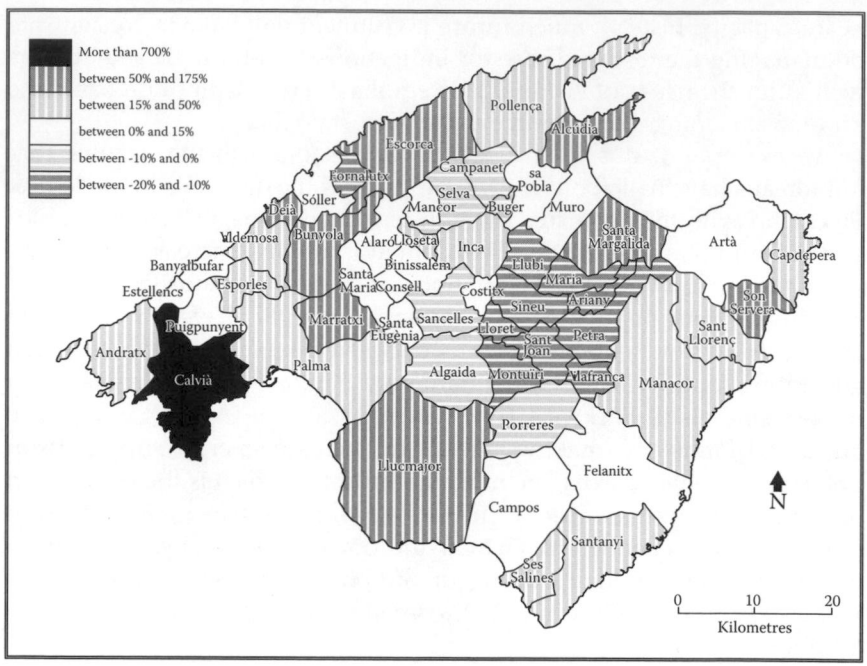

Figure 6.1 Map of population change from 1970 to 1996
Source: Llibre Blanc (2009)

arriving in the seven years from 1999 (Llibre Blanc, 2009: 220–221). The increase in a single year – 2007–2008 – was even more remarkable when the population of the Islands increased by almost 4%, and so now nearly 21% of the population is overseas born, this group increasing by 17% in that year alone! Professor Pere Salvà believes that the actual figure (registered and non-registered – illegal?) is nearer 25%. Only about 50% of the Spanish population was born in the islands. Of the foreign born, Germans are in a majority (33,000) with Moroccans second (20,000) and Latin American born third, including 13,000 Ecuadorians, 12,000 Argentineans and 10,000 Columbians (INE and Diario de Mallorca, 21 June 2008). These data are for the whole of the Balearic Islands although nearly all of these changes have been concentrated in Mallorca. Birth rates and natural change have been increasing particularly amongst mothers born overseas. In 2001, 14% of all island births were in this category; by 2008, this had risen to 32% (Social and Economic Report of Balearic Islands, 2008: 279). It can be postulated that much of this process is concentrated amongst immigrants from Africa and Latin America as many North European immigrants are likely to be above child-bearing age.

In addition to this resident population, the transient or tourist population needs to be considered. We have seen that tourist numbers in Mallorca have risen from 5,874,500 in 1994 to 9,631,100 in 2008. We know that there is a marked seasonality to these inward flows of foreign visitors, with about 72% visiting in the high season (May to September); tourists from the Peninsula show a slightly flatter distribution (see Figure 2.2).

The problem, as part of the impact of population, is to derive a meaningful figure of tourists that can be added to the resident population to give an indication of total or what might be called the 'real' population. All the population, whether transient or resident, has an environmental impact; earlier demographic analyses in Mallorca tended to include only registered residents in which the non-Spanish element usually seemed an underestimate; tourists were excluded. When these figures were used for planning purposes, they were of limited value. However, simply adding the 2008 figure of 9.6 million tourists to 0.8 million residents (10.4 million) has little meaning. For the high season, adding 72% of 9.6 million (6.9 million) to 0.8 million residents (7.7 million) has more meaning. It would give an average density of 2856 per km^2, an equally unhelpful figure.

Clearly, for environmental purposes, there is a need to be better able to calculate the real or actual population on the island at any one time, that is, the registered residents plus recorded tourists and other visitors but not including unknown illegal immigrants. To this end, Mallorcan scientists have devised a useful measure of population that can be used when considering the environmental pressures involved. This is *l'indicador diari de pressió humana* (IDPH), a measure of daily population pressure (Llibre Blanc, 2009: 225). It is based on earlier work by Blázquez *et al.* (2002: 48–65), which showed that the average real population (IPH) for 1999 was 884,894 but for August this rose to 995,244, an 11% increase. Later, the Llibre Blanc (2009) calculated that on average a weighting of 27% above the registered resident population was required to record the real average demographic pressure over the period 1998–2006. By 2008, the maximum IDPH was reached on 8 August when the 'real' population of the Balearic Islands reached 1,819,606 or 18% above the registered resident population. The IDPH ranged from 51.3% above for the high season (60.9% for August) and gave a figure close to the actual resident population for the low season. It has been calculated that for Mallorca the maximum of 1,316,251 was reached on 8 August 2008 and the minimum of 830,535 on 1 January 2008 (Social and Economic Report of Balearic Islands, 2008: 427). Using these methods, it is possible to show the seasonal variations in population pressure resulting from tourism and hence its possible environmental impact.

Another, but perhaps less robust, technique that has been deployed in recent years has been to examine the hypothetical impact of an individual tourist via what is called the ecological footprint. This is a means of measuring the impact of one tourist per day on the environment based on resources used and waste generated, and then converting it into an areal measure. For example, how many hectares of woodland would be needed to absorb CO_2 emissions of one car? It attempts to calculate the hypothetical area that would then be required to reduce the impact to zero. The value of this technique is questionable since most efforts have been based on global averages. Estavan and Llorente (2001) calculated that for Mallorca's actual 364,100 ha it would need a footprint of 4,370,000 ha to accommodate the impact of its real population (residents plus tourists); that is, it has an ecological deficit equal in area to that of Japan or Holland! If tourists represent the equivalent of 228,000 residents in environmental impact terms, then their footprint is equal to 5.3 'Mallorcas' while registered residents and tourists are equivalent to 6.7 'Mallorcas' (Estavan & Llorente, 2001: 90).

A more traditional measure of environmental impact but now given little credence is carrying capacity, which the World Tourism Organization defines as 'the maximum number of people who can visit the same place at the same time without damaging the physical, economic and socio-cultural environment, without reducing the quality of the visitor experience', a useful notion but one notoriously difficult to quantify. It has a number of dimensions from the physical through the socio-economic and cultural. Its initial borrowing from ecology gave it some support in early years in Mallorca, but because of this complexity it has been little used in practice. Calculating the number of visitors that an area can support will vary in many ways, not least with the nature of their activity as consumers and the duration of their stay. In Mallorca, the political response to increasing numbers has been to rely on supply-side technological solutions – increasing the uptake of water supplies, reservoir construction, building additional sewage works, artificially rejuvenating beaches, motorway construction, expanding the rail system and above all, increasing imports. The list is an endless litany of the process of modernisation, some elements of which are examined below.

The key question then becomes the extent to which this process can be continued without environmental collapse. 'Damage' has been by-and-large politically acceptable to many, with the price of environmental degradation seen as worth paying to allow the almost exponential growth in tourist and resident numbers to continue because of the net wealth generated. However, public opinion on the possibilities of continuing technological solutions has shifted remarkably in Mallorca in recent years, and political pressure for containing or ameliorating the effects of the tourism industry – even reversing them – have grown. This leads us naturally to a consideration of sustainability as an alternative construct.

Sustainability

Public opinion of both local residents and tourists reflects the need for better management of the impact of tourism on the Mallorcan environment. The difficulty of shifting pressure from certain areas at certain times of the year without reducing the conventional economic benefits is equally recognised; it appears to be an intractable problem. There is strong political and economic pressure not to kill *'la gallina dels ous d'or'* (The goose that lays the golden egg), something recognised even in the Franco era (Buades, 2004:138). Searching for a solution to the apparent contradictions between economic growth and environmental protection, Mallorca like so many other tourist regions – and following the example of European Union's policies contained in, *inter alia*, the 1986 Single Europe Act and the treaties of Maastricht (1992), Amsterdam (1997) and Nice (2000) – appears to have put its faith in the notion of sustainability (Baker, 2006:136). As economies such as Mallorca's have moved to be dominated by the tertiary and quaternary sectors, new technological solutions and changes in resource use have led to new ways and means for state intervention. Reducing risk becomes a central activity through management and planning, but, as Baker reminds us, the 'old' obsession with pollution and resource depletion now has to be complemented by the social and cultural changes that must be part of sustainability policies. Once criticised as being resource and pollution obsessed to the exclusion of the social and cultural changes necessary to promote sustainability, consumption now has to be addressed. Market solutions and managing consumer behaviour are more likely to be relevant than state dictat (Baker, 2006: 138–139). Put simply, this means directing efforts towards modifying tourists' behaviour towards the Mallorcan environment.

This is a complex topic not really susceptible to its popular adoption by politicians and the media as that doyen of tourism academics Richard Butler pointed out in a critical chapter 10 years ago. He reminded his readers that the term had already been in use for 10 years (now 20 years ago), with its origins in the preservation and conservation movements of the late 19th and early 20th centuries and in the Limits to Growth debate of the early 1970s. Sympathy for its goals has not necessarily translated into 'acceptable costs and sacrifices that actual application may entail' (Butler, 1998: 26). What those who have too readily adopted the term, and even its proper scientific application, have often failed to realise is its holistic nature and the difficulties of separating tourism from practically all other aspects of socio-economic life. The view taken by Bramwell (2004: 17) is that 'sustainable development is a "socially constructed" and contested concept that reflects the interests of those involved'. For Mallorca, sustainability posits something more than simple 'costs versus benefits', involving contested discourses between many elements of island society. When one industry – tourism – dominates the economy and a large section of cultural life, the choices are much starker than in a more diversified system whether that be economic or ecological (Mayol & Machado, 1992: 81ff). Having invested in what some see as a monoculture, any action will have to be undertaken cautiously and over a long period of time. However, rapidly changing external events such as a world recession or climate change over which the island has no control may force change, the results of which are unpredictable.

Meanwhile, sustainable tourism is not achievable outside the sustainability of the island economy as a whole. It may remain a political objective but a better one might be to aim for impact reduction, that is, taking those steps that might reduce the negative effects of tourism on Mallorca's environment. This view accords with Buckley's definition that 'sustainable tourism means tourism at any scale with practical and proactive design, engineering and management to reduce environmental impacts' (Buckley, 2002:184). Unfortunately, as Butler pointed out, '"sustainability" has been adopted by the tourism industry for three reasons: economics, public relations and marketing. It has become hijacked as a moral-laden term to improve the respectability of mass tourism; in practice its effects have been minimal' (Butler, 1998: 27). For mature economies such as Mallorca, using sustainability as a tool to help improve the quality of the tourism product is difficult. In some senses, Mallorcan tourism is rather like a supertanker, slow and difficult to turn around. On the other hand, there have been successes in the field of impact minimisation in Mallorca that can be pointed to but public perception of these is limited. In the case of Calvià municipality, for example, we find the ironic juxtaposition of the area of Mallorca most densely populated with 'mass' tourists and a local authority that has tried to do more than most to ameliorate the effects of environmental degradation. In 1960, Calvià had 3000 residents and 6800 bedspaces in 112 tourist establishments. By 1997, there were 35,000 registered residents, 50,000 de facto residents, 120,000 bedspaces and 256 tourist establishments. It receives nearly 1.6 million visitors annually, 20% of Mallorca's total who occupy the area's 27 beaches spread over 54 km of coastline. It is hardly surprising that it had become characterised as an area of 'badly co-ordinated development, unlimited construction, growth and unsustainable use of natural resources' (Dodds, 2007: 300). As early as 1995, however, it adopted Agenda 21 following the Rio Summit and used local planning regulations to control construction and improve environmental quality. Aguiló *et al.* have been able to show that some of the 3% higher prices charged for package holidays in that municipality are attributable to these improvements in quality (Aguiló *et al.*, 2005: 228–229; Dodds, 2008; Nájera & Bustamante, 2007).

The proper measurement of sustainability is the first stage to policy implementation. To this end, the Govern Balear has followed Calvià's pioneering efforts and recently approved the widespread adoption of the Agenda 21 principles originally set out in the Aalborg Charter (1994) and its further Commitments of 2004. A set of 98 sustainability indicators has been identified, which includes aspects of tourism's effects, and data are now being collected on a municipal basis so that temporal and spatial measurements will be available shortly. Mallorca presents particular problems because of its size and complexity compared with the other Balearic islands (Govern Balear, 2010).

A second weakness of the use of the term in tourism studies is that any improvements often remain largely invisible to the short-term tourist. Perhaps only direct consumer shortages or interruptions of supply, such as water and electricity supply problems, affect decision making about satisfaction or a decision to return (Aguiló *et al.*, 2005: 227). Mallorcans may have recognised the tension between making changes to personal and societal behaviour on the one hand and technological solutions on the other, but since tourism is essentially a 'people process', technical solutions are often conveniently more acceptable politically to both the host and the visitor. The local population makes its living from mass tourism and seeks to minimise costs and maximise benefits, and holidaymakers, for the most part, do not want to be concerned with their environmental effects during their short stay on the island. As long as technological solutions can keep the environmental dangers at bay, or hidden, the tourists will keep returning to what they perceive to be a constant environment. In the economists' favourite encomium – the long run – this could change, but how many tourists' visits does it take to make up the long run? In the long-term growth curve of Mallorca's tourist numbers, we have identified what proved to be minor reductions or halts in growth – 1973–1974 and 1991–1992 – and now in the global recession of 2007 onwards – but all of these are attributable to exogenous factors. It would seem that if the Mallorcan tourism environment is deteriorating – and parts of it certainly are – then it is difficult to prove that it does not affect tourist demand. Could this be because tourists from abroad have become inured or insensitive to such decline from their experiences at home? Could it be that in relative terms Mallorca still appears 'better' than home at least for two weeks in the summer? Mallorca may no longer be the paradise that it was to earlier visitors, but it is still perceived to be an attractive venue.

Spatial and Temporal Variations

Two dimensions to the population aspect of environmental impact have to be considered – the spatial and the temporal with both dimensions having a long-term (historical) aspect and, in any one year, a seasonal one. The spatial dimension is important because population has become highly concentrated in the traditional urban centres, especially in Palma, which historically has always contained between 25% and 50% of the island's population, and in the urbanisations that have grown up over the last 50 years as the locus of the coastal tourism industry (Barceló Pons, 1970; Salvà Tomàs, 1990). Two further spatial processes have been at work, one centrifugal, whereby the heart of the island suffered considerable population loss during the 'boom' periods of tourism (1955–1985), and a centripetal one, which has seen a counterurbanisation process, especially in the south of the island, in which population has been decentralising from the capital city and now supports self-sustaining growth in and beyond the wider metropolitan area (Binimelis, 1998: 198). The density of population naturally rises over time since Mallorca is an island and its land area may be taken as a constant (Figure 6.1).

What kind of environmental pressure do these aspects of population dynamics present? Irrespective of temporal and spatial variations, the level of increase described above has made enormous demands on three resources in particular: water, energy and land. Naturally, the per capita consumption of these three has increased as Mallorcan society and that of its visitors has become more affluent since the 1960s.One simple example of that is the installation of air conditioning equipment in homes, hotels, offices and workshops, increasing electricity consumption. A second mentioned above has been the increase in car ownership by residents and car-hire use by visitors. Today, few homes in Mallorca are satisfied with one bathroom and so water use has also increased; since the 1980s, nearly all hotels have had en-suite facilities. Land uses have changed as resorts were built and as all settlements expanded at their edges. Building in the countryside has especially grown in the last decade. So, total consumption is a function of increasing numbers of consumers and increasing affluence and expectations.

Tourism in Mallorca, as we have shown in Chapter 4, shows marked spatial variations in its intensity; it is essentially coastal with particular concentrations east and west of, and including, Palma, much of the east coast and the two northern bays (Figure 6.2). Of Mallorca's 285,370 tourist places in 2008, these three zones contained more than 90% of them: 36% in the south, 32.6% in the north and 22.2% in the east. Only the cliffed west coast has relatively few tourists. Even within these three zones, certain concentrations are noteworthy. On the east coast, the

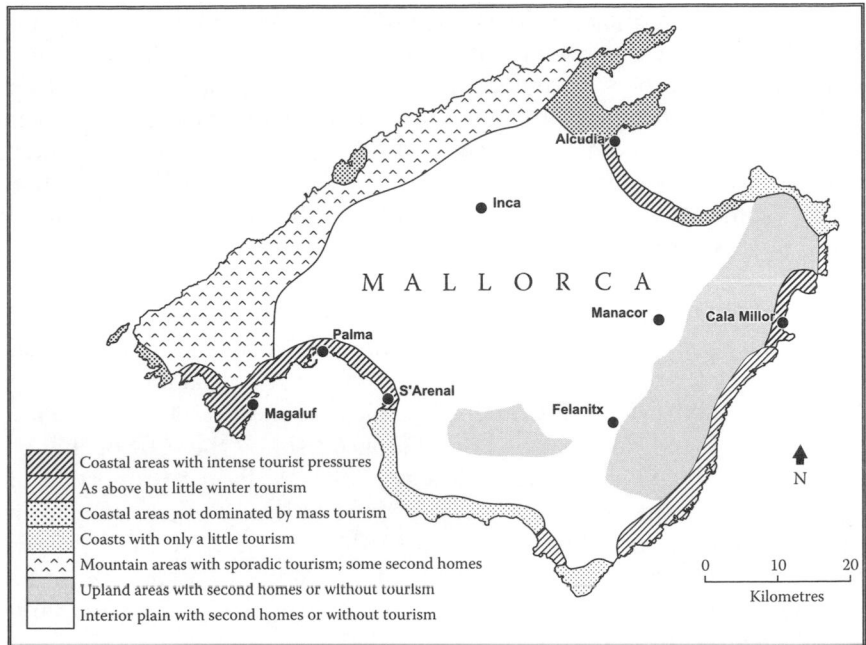

Figure 6.2 Map of tourism pressure
Source: Llibre Blanc (1987)

almost continuous 'wall of cement' that consists of Cala Bona, Cala Millor, Sa Coma and S'Illot contains nearly 13% of all of Mallorca's places; Calvià municipality alone in the southeast has an even greater concentration. Palma with resorts to the east and west of the city centre has 15.3%; if amalgamated with nearby S'Arenal to the east, this figure rises to 19.7%. This spatial concentration is, then, remarkably linear along the coast, with one resort merging imperceptibly with its neighbour, the result of adjacent municipalities seeking development of their own section of the coast, the classic example being Cala Bona and Cala Millor being 'shared' between Sant Llorenç and Son Servera municipalities (Buswell, 1996: 330).

When the environmental impact of mass tourism began to be felt first in the 1960s, the principal concern appeared to be for the island's flora and fauna and associated ecosystems. This was because early research was led by biologists (Mayol & Machado, 1993). An early pressure group was GOB (*Grup Balear d'Ornitologia i Defensa de la Naturalesa*), established in 1973, initially a pro birdlife lobby with a large membership but which has since become concerned with all aspects of the environment, especially in recent years with road-building programmes (www.gobmallorca.com). As is often the case, early pressures were for the protection of certain species, especially those endemic to the island. Where these plants and animals were part of a coastal ecosystem, this seemed understandable and rational. Most of the early environmental legislation, then, was to set up protection for valued 'natural' areas. The first law of Natural Spaces was passed in 1991 with the second one in 1994, whereby 34% of the islands' territory was theoretically protected (Picornell & Picornell, 2002a:104). Today, some 18.7% of Mallorca's territory is under some form of protective legislation.

However, the main impact was on the human ecology; mass tourism produced a totally new urban environment in locations formerly untouched by much building. What became the focus of environmental opposition to tourism was the deleterious effects of coastal town building (Blázquez Salom, 2005). For many, the major issue has been the rising proportion of land classified as urban. Table 6.1 shows the scale of change as calculated by Murray *et al.* following Pons. While the proportion of urban land appears relatively low, it is more in line with experiences in northern and western Europe than in Mediterranean countries. Clearly, the conversion has been from agricultural land to urban uses; the proportion of hard-to-define 'natural' land has remained fairly constant.

With tourist places in the Balearic Islands increasing by 34% from 312,050 in 1982 to 475,056 in 1988, but with the non-regulated part of this rising from 45,629 to 135,927, that is, by 66%, it was not surprising that the early legislation began to shift towards building control. The first law of non-hotel accommodation was passed in 1984 to try to control these illegal apartments. By 2000, as much as 35% of the accommodation capacity was in this category, but more recently the emphasis in building has moved to residential housing in urban and rural locations. While the second home or the holiday villa for letting may be set in attractive rural environments for marketing purposes, their presence has been slowly transforming the countryside from an appearance based on farming to one that is much more urban in character, an example of *urbs in rure*.

Table 6.1 Balearic Islands' land-use changes from 1950s to 2000

	Urban land use (%)	*Agricultural land use (%)*	*Natural environment (%)*	*Total (%)*
1956	1.18	61.3	37.52	99.8
1973	2.54	60.2	37.98	100.72
1993	5.15	58.07	37.50	100.72
2000	5.44	57.86	37.41	100.51

Source: Murray *et al.* (2005) and Pons (2002, 2004)

Clearly, most of this building in the countryside is for tourist or second home ownership, with local people making money from the sale of underused agricultural land and from construction. Local communities are unable to meet the prices asked for such properties nor are they designed with such people in mind. However, as Binimelis has shown, the process of urbanisation of the countryside is by no means confined to the tourist incomers (Binimelis, 2002). More affluent city and town dwellers in Mallorca have been decanting into the countryside for some time. Much of this villa studding has been done without proper planning permission. Blázquez Salom points out the while property tax registrations in Mallorca rose from 4000 in the mid-1970s to nearly 39,000 25 years later, only 7000 building permits had been allocated in that time (Blázquez Salom, 2005).

The first of the Cladera Laws was enacted in 1984 limiting new hotel construction to 30 square metres per bed space, a figure raised to 60 square metres in 1988 together with a limit on the number of three-star hotels that could be built. By the 1990s, the rate of illegal apartment building while far from being eradicated was at least slowing down, with only a 2% increase in places from 1993 to 1999 (Picornell & Picornell, 2002b: 97). In addition to the effects of these illegal buildings, the other effects related to construction and the sheer numbers of tourists seemed to be given much less prominence in the legislation (Williams & Shaw, 1998: 55). Sewage disposal, waste management, water supply, the 'mining' of the landscape for building materials, the surface changes wrought by road building, the impact of rising car ownership, the damage to the lower reaches of the surface systems and to 'torrentes' leading to flooding, the rising incidence of wildfires especially near built-up areas – this almost endless list did not immediately result in a move to environmental management before the mid-1980s perhaps because it presented more difficult challenges to Mallorcan authorities than the protection of flora and fauna and landscapes.

In any consideration of tourism, seasonality is important as has often been stated. With the pressure of transient population concentrated into 4 months of the year (in effect in two – July and August), the demand for many resources is heavily unbalanced. Power and energy, water, food supplies and labour are all in greater demand in these months and so supplies have to be increased to very high levels, creating problems of generation, storage and distribution. In economic terms, the island has to provide an infrastructure that is used only temporarily: reservoirs (there are only two on Mallorca, Gorg Blau and Cuber, both built in the late 1960s in response to tourist demand) have to be large enough to meet peak demand as do the electricity generating stations. Food supply in Mallorca in the high season can be met only by a massive increase in imports. The external costs of this supply system are by-and-large met by the population as a whole not by tourists alone. Similarly, the environmental costs (externalities) of meeting high seasonal needs, especially those created by pollution and waste disposal, have to be paid for by the public sector, that is, permanent residents supporting tourists. Clearly, some kind of redistributive mechanism was, and is, necessary to improve equity and social justice for local residents: the seasonally excessive consumer and polluter ought to be made to meet its share of these costs.

Technical Problems and Partial Solutions

In the recently published Llibre Blanc (2009), the authors identified six areas of longstanding environmental concern thought to be capable of improvement via technical solution (Fortuny *et al.*, 2008; Llibre Blanc, 2009: 238–282). Most of them are the product of excessive human behaviour and are all susceptible to impact reduction processes. Three are worth examining in more detail: water supply, the disposal of waste products and energy and pollution.

The Ministry of the Environment in the Balearic Islands has estimated that tourism consumes about 40 hm^3 of water annually, but agriculture remains the principal consumer, which raises the question as to whether or not tourist consumption can really be a constraint on the development of tourism (Essex *et al.*, 2004). The relatively short peak period when tourist consumption exceeds the residents' figure – early June to mid September, with a peak in August – requires 160% above the mean consumption (Figure 6.3). Expensive infrastructure is required to meet this seasonal demand, but the cost has to be borne by residents or through general taxation. Eighty percent of water supply comes from subterranean aquifers, but many of the concentrations of tourist urbanisations around the south, east and north coasts draw on local aquifers, lowering the water table in the summer months and allowing salt water ingress; this now affects 39% of the underground sources. Forty-three of the 90 underground sources in the islands are now contaminated to a greater or lesser degree by fertilisers, sewage leakage and other pollutants. Only 2.5% of water supply in Mallorca comes from the two reservoirs, Cuber and Gorge Blau. We have already noted the consumption of grey or recycled water being compulsory for the island's 20 golf courses (4.8 hm^3 in 2006), and this constraint is now applied to agricultural irrigation where practicable (18.0 hm^3) and parks and gardens (1.8 hm^3). There have also been experiments to use such supplies to flush hotel toilets (Gual *et al.*, 2008). Water from such sources is not suitable for human consumption; hence, the island has resorted since the late 1990s to very expensive desalination systems drawing water from contaminated subterranean sources and directly from the sea (http://www.watertime.net/docs/WP2/D35_Palma_de_Mallorca.doc).

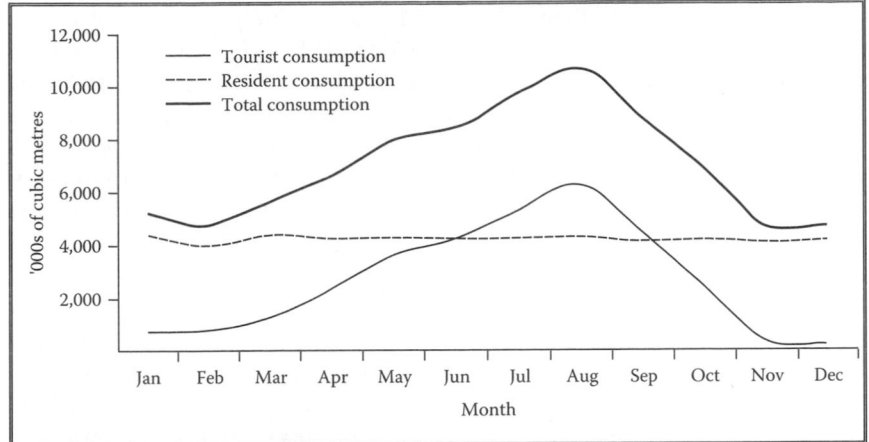

Figure 6.3 Water consumption by tourists and residents
Source: Llibre Blanc (2009)

The costs are, however, prohibitive. Subterranean water can be produced for between 10 and 15 centimos per cubic metre, but desalinated sea water and purified water cost about 1 euro per cubic metre (Rodriguez Perea, 2010). Occasionally (1995–1997), Mallorca has resorted to importing water in tankers from the Ebro Basin on the Peninsula, but this has not proved a successful technical solution.

When nearly 9 million tourists descend on an island of 3641 km^2 in a year, the disposal of waste and its management naturally becomes a challenge. If landfill were to be the only means of disposal of solid waste, then a small island would soon exhaust its limited sites for burial. Excavations are only part of the problem as most landfills may also contaminate soils, water sources and the atmosphere. However, tourism per se is primarily responsible for large amounts of traditional urban waste, which when combined with residents' waste gives an annual total of 732,000 tons of food, paper and cardboard, glass, cans etc. Added to this are 520,000 tons of industrial waste from manufacturing industry and residues from construction and demolition (much of it indirectly related to tourism), giving a total of 1,252,000 tons for Mallorca in 2008 (Consell de Mallorca, 2009). In 2006, per capita urban waste generation for disposal in the Balearic Islands was 602 kg, high by Spanish standards (484 kg/capita) and even other tourist regions such as Andalucía (499 kg/capita). It has been estimated that the tourist population in the third quarter of the year adds about 31% to Mallorca's total urban waste. Again, geographical distribution of waste production is important and so the resort municipalities add considerably more – Calvià 135% and Deià a staggering 280% increase. Calvià's urban waste in August is double that in January (Llibre Blanc, 2009: 254). Two principal means of urban waste disposal are practiced on Mallorca: landfill and incineration, which account for 33% and 42%, respectively, while composting (18%) and recycling (8%) make up the remainder. While the three 'Rs' of waste management (reuse, recycle and repair) have been fairly successful in reducing the amount for disposal, tourists are difficult to include in any recycling policy. Only through their hotels may something substantial be achieved. In any case, holidays by definition are consumer orientated and additional, often gross, consumption is an unwritten objective of many tourists. The long-run objective is to try to reduce landfill to zero and to raise recycling to 40% of the total (Consell, 2009), but how this might be achieved in a mass tourism economy is difficult to foresee.

Energy supply presents particular problems to a tourist region such as Mallorca that is dominated by (1) seasonality and (2) massive temporary migrations. Historically, the island depended upon water and wind power to produce energy for industry and manufacturing. Domestic energy came primarily from charcoal for heating purposes (Buswell, forthcoming). The introduction of steam power in the late 19th century depended upon coal imports. It was only with the introduction of electricity and the internal combustion engine in the 20th century that Mallorca was able to modernise energy use and experience the associated growing problem of pollution, accentuated by the onset of mass tourism and urbanisation from the 1950s.

The traditional solution to energy shortages was to increase supply by constructing new power stations or by increasing the efficiency of electricity distribution by overhead lines; only Palma originally had town gas, the remainder of the island relying on subsidised bottled butane and propane distribution by road. As tourist numbers rose in the first and second 'booms', the islands' electricity consumption inevitably rose too, from 496 KWh in 1965 to 2500 KWh in 1988; in the five years to 2005, it increased by another 27.5% and is likely to increase annually at a rate ranging between 4% and 6%. It was clear that new capacity was needed. Seasonality is yet again a dominant characteristic of consumption, with 46% of all petroleum product sales and 38.2% of the electricity generated taking place between June and September (Llibre Blanc, 2009: 247). Local generation was seen as yet another contributor to pollution and at the same time it was not able to develop economies of scale. Part of the solution has been to construct an undersea natural gas pipeline from the mainland taking advantage of national bulk supplies. The new combined cycle plant – Son Reus II – has increased generation capacity on the island by 18.5% and given the island (and Menorca with which it is linked) a useful reserve. In sustainability terms, this may give a generating system that meets national and European standards, produces less CO_2 and contributes less to global warming, but in reality it moves much of the problem elsewhere. An added advantage of the new natural gas source is that an island-wide network of distributing pipelines is being constructed for domestic supply starting in a hierarchical fashion with the larger towns first.

To complement these 'improvements' in energy supply, the island was linked by undersea DC cable to the Peninsula's national grid in February 2011 (Diario de Mallorca, 7 January 2011) and will start operating early in 2012. While this will no doubt safeguard security of electricity supply, help offset Mallorca's high seasonal demand and reduce Mallorca's pollution from generation, it will be another example of nimbyism.

Getting There and Getting About

Transport is a major contributor to both the quality of life and environmental degradation. For tourists, it enables them to travel long distances to reach their destinations at relatively low costs to themselves but imposes hidden costs on the environment that they do not have to meet directly. For Mallorca, two aspects of this apparent contradiction may be considered: travel to and from the island, and travel while there.

The principal means of movement for tourists en masse, and to an island, is clearly by air; in earlier chapters we highlighted the role of air transport in 'inventing' mass tourism, increasingly so from the late 1940s onwards. Some basic data illustrate the scale of movements to and from Mallorca. In 2009, there were 177,502 flights in and out of Son Sant Joan airport carrying 21.2 million passengers and 17.1 million tons of goods. In the year immediately preceding the recession, these figures were 197,384, 23.2 and 22.8, respectively, (http://estdisticas.aena.es, accessed 4 April 2010). It is not possible to calculate accurately the impact of these movements on Mallorcan, or even national, airspace but the airport is now the third largest in Spain and 10th in the European Union but later in this chapter we will consider the role of air transport in climate change. Besides CO_2 there is now concern for other gaseous emissions such as nitrogen oxides and their impact on global warming. The impact on local air quality in the vicinity of the airport in Mallorca is more readily measured as are factors such as the pollution by the airport and its activities in servicing planes and processing passenger throughput. Movement of passengers by ferry is less than a quarter of a million, 85% of them Spanish. In addition, more than 700 cruise ships dock annually in Palma. Pollution per passenger kilometre by sea is now reckoned to be a larger contributor than originally thought and sea transport now figures more prominently in debates about sustainability. A ferry, for example, emits about 120 g CO_2 per passenger/kilometre; a cruise liner on average emits 401 g per passenger/kilometre, the latter approximately three times that for a modern airliner (*The Daily Telegraph*, 19 January 2008).

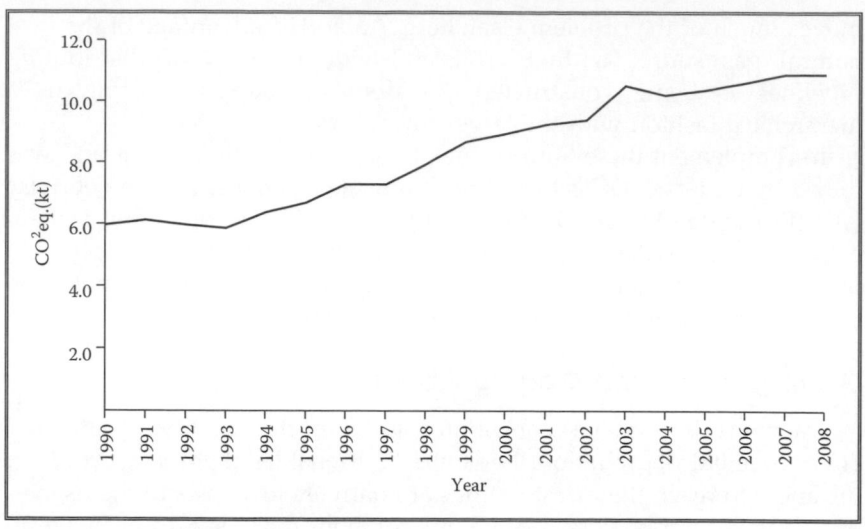

Figure 6.4 Evolution of greenhouse gases from 1990 to 2008
Source: Govern Balear (2008)

As with aircraft movements, allocating Mallorca's share from shipping to its contribution to CO_2 emissions is difficult to calculate but more ferries/cruise liners must equal net additional global pollution from tourism (Figure 6.4).

In addition to movements to and from the island, a second item is a consideration of the impact of the movement of tourists within the island. This takes three forms: coach and taxi travel between airport and resort or hotel; the use of coaches for side tours and car hire. Tourists' use of the rail system in Mallorca is at present limited; it was not designed in the 19th century to link ports with coastal resorts. Their use of local public transport by bus is mostly restricted to the south coast for journeys to and from Palma.

Car ownership by Mallorcans is amongst the highest in Spain, a product of the affluence produced by tourism. In the most recent survey (2001), there were 738,362 vehicles registered in the Balearic Islands of which 562,855 were cars, giving 877 vehicles per 1000 head of population or 699 cars per 1000 head of population at that time, figures that have almost certainly increased in the last decade. If tourist population numbers are included, the figure for cars falls slightly to 524, but the fact remains that there are at least 40% more cars than registered drivers. Mallorcan society is highly mobile, making 2.19 journeys each day (Madrid 2.31) and two thirds of these are by motorised transport, 87% by private car (Madrid 56.1%) with only 13% by public transport (www.caib.es/dgtransp/esdev/pla/avaluacio.html, accessed 15 March 2010).

Table 6.2 Balearic Islands' pollutants

	Pollutants				
	SOx (t)	*NOx (t)*	*CO (t)*	*CO₂ (kt)*	*PM10 (t)*
Transport by road	60	9,563	11,544	2,372	659
Other transport	12,509	18,690	3,159	1,622	2125
Total all sectors	26,557 (47.3%)	53,792 (52.5%)	26,699.9 (55.1%)	10,991 (36.3%)	3963 (70.2%)

Source: Govern Balear (2008)

Car hire is highly significant in Mallorca and the islands as visitors seek to replicate the convenience and symbolism of 'ownership' that they undoubtedly exhibit at home. In 2010, about 38,000 hire cars were located in the islands, more than 30,000 in Mallorca, perhaps the greatest concentration in Spain. Even so, this represents a 35% decrease over 2009 but prices to hirers rose by 20%. It has been estimated that turnover of the car hire business in the Balearic Islands in 2009 was 144 million euros (Ultima Hora, 11 May 2010).

These data, especially for car ownership and use by Mallorcans and visitors alike, demonstrate the considerable impact of transport on the local environment and on imported supplies of fuel. As in other developed economies, it is transport in particular that is making a considerable contribution to pollution as Table 6.2 – and the next section – shows.

The Sant Son Joan airport is now a major consumer of land as are the berths and docking facilities of the Palma's port and the high-speed roads that connect them both to the major city and its suburbs. Transport's environmental effects are by no means confined to greenhouse gases. Note too the spatial concentration of these transport systems in the south central part of the island, part of the megalapolitan effect of Greater Palma and its city region.

Tourism and Climate Change

The period since late 2007 has been characterised by two apparent crises: the global economic downturn, which has been particularly acute in Western Europe, including Spain, and the revelations concerning global warming that were further exposed as a major policy issue for the world at Copenhagen in 2009. The impact of the former will be considered vis-à-vis Mallorca in the book's final chapter, but since 'climate' is a defining factor for Mediterranean tourism, it will be considered here; any long-term changes may have serious implications for the Balearic Islands.

Any consideration involves a number of aspects that are by no means solely associated with what happens in Mallorca itself; the world's climate has to be accepted as a global system. What happens to climate (and weather) in Mallorca is the product of mainly exogenous forces largely because it is a small island with a small population and so the global impact of indigenous forces is small but compounded by tourist travel, population and consumption.

Transport gives considerable cause for concern in the arguments about sustainability. The Islands' emissions of greenhouse gases (GHGs) have been rising steadily as shown in Figure 6.4.

It has been estimated that 36% of Mallorca's CO_2 emissions are derived from transport, 58% of which is from road transport. It is difficult, however, to calculate how much of this is derived from tourist activity, but in many ways outputs from residents should not be separated from those from tourists: the atmosphere does not distinguish between them (see Table 6.2; changes in rainfall and temperature since the mid-1950s are shown in Figure 6.5).

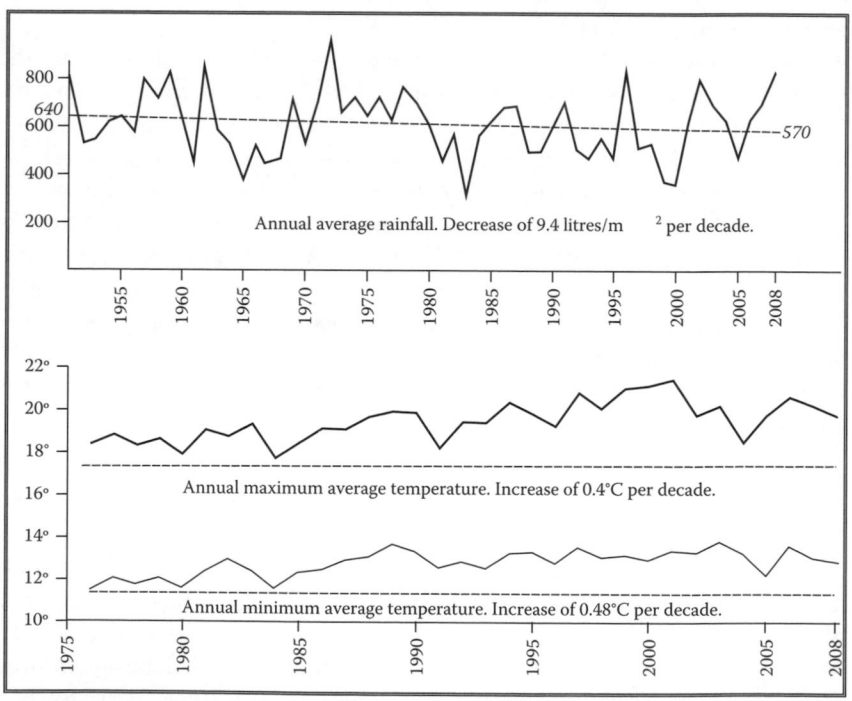

Figure 6.5 Climate change (rainfall and temperature) in the Balearic Islands
Source: Diario de Mallorca (3 October 2009)

However, estimates have been made of the output of CO_2 by tourists in Mallorca (Travel Foundation, 2007; WWF-UK, 2002). Few methodologies are very robust for calculating such impacts. The two reports consulted were based on 2002 data and used British DEFRA methodology for drawing comparisons with outputs at home. Some of the variables that have to be taken into account besides simple numbers of tourists include length of stay, type of accommodation, travel to and from the island and within Mallorca including car hire, food and water consumption and processing waste. The principal means of measuring these is through energy consumption and its production of CO_2. Only the impact of British visitors is considered here; results from each country sending tourists would be different.

With these caveats in mind, these reports found that the CO_2 emission for UK visitors in 2002 was 103,000 tons, or 508,000 tons if flights were included. This leads to a rough estimate of the carbon emission per bed night in Mallorca as 27 kg. This can be compared with the average domestic per capita energy consumption in the United Kingdom of about 5.4–6.0 kg CO_2 per day. According to the calculations made in these reports, about 14% of the CO_2 emitted was derived from hotel energy consumption, 6.2% from transport used within the island and a startling 80% from the flights to and from Mallorca. These figures bear comparison with similar findings in Australia, for example, where total transport activities accounted for 82.2% of all GHGs emissions, accommodation accounted for 4.5%, other support activities 8.6% and retail 3.4% (Forsyth *et al.*, 2007).

It is hardly surprising that the effect of the tourism industry's air transport on atmospheric pollution has begun to receive a great deal of attention, curiously little of it originating in Mallorca. Today, only about 17% of all trips worldwide are for touristic purposes, but their contribution to CO_2 emissions is disproportionate. Of all tourist trips globally in 2005, 45% were by air, but this figure is predicted to rise to 55% by 2035, faster than air movements generally. For tourists in the European Union, while 20% of tourist trips are by air, over half of all passenger/kilometres are attributable to tourism. Aircraft contributed 5% of all CO_2 emissions at the beginning of this century, but this proportion has already risen to 8%–10%, partly the result of more flights but partly because of falling emissions from other sectors where government policies are beginning to have an effect (Bows *et al.*, 2009: 16–17); 17% of all global air movements could be accounted for by tourism in 2005, and this is likely to increase to 26% by 2035 (Bows *et al.*, 2009: 17). Mallorca is reliant upon the movement of tourists each year by air, more than 9 million in 2009, and attention in recent years has focussed on the relative impacts of overseas transport on climate change. Estimates vary, but by 2050 at least 5%–10% of the total GHG effects on climate change will emanate from air transport, which is already the fastest growing contributor to GHGs (Ruijgrok & van Paassen, 2005). Clearly, Mallorca's tourists may have a considerable impact on the *local* environment in many ways once they have arrived, as we have seen, but their travel patterns to and from the island have a much larger impact on the wider regional and, in turn, global climatic environment. Technical efficiencies in design are making aircraft more fuel efficient and therefore less polluting, but these latest versions are slow to be adopted on short-haul European routes and therefore they are unlikely to offset the effects of the projected growth of arrival by air to Mallorca (Bows *et al.*, 2009: 10). This is compounded by the plans of island authorities to expand further Son Sant Joan airport. Substituting air travel for land and sea travel is a cloud cuckoo prospect.

For tourism, climate change may be examined under two sets of headings: the climate itself (and its associated weather patterns) and its effects on other factors notably water supply and drought, the incidence of forest and other fires, health and sea temperatures and levels.

There is a growing literature on the possible effects of climate change on tourism and tourists that cannot be reviewed here (but see Bigano *et al.*, 2005; Davos 2008; the Djerba Declaration from the WTO, 2003; Viner & Agnew, 1999, for example). For tourism, two aspects of change in the actual climate are important: shifts in temperature and changing rainfall patterns. Any scrutiny of these will be based on the predictive power of current models to forecast levels at various points in the future. The Mediterranean region is expected to see temperature rises of 0.3°C to 0.7°C with an increase in the number of days over 40°C, a decrease in summer rainfall of about 15% and an increase in winter rainfall with more of it in intensive storms. The possible effects of these climatic changes on tourist flows from northern Europe to the Mediterranean were usefully summarised by the Djerba Conference (WTO, 2003):

> *Broadly, this is a market with a single purpose. It leaves behind an unpredictable summer climate in northern Europe – possibly little sun, plenty of rain and cool temperatures – in search of an annual dose of certain warmth and sunshine. Altered weather patterns induced by climate change could mean that northern Europe becomes more attractive and reliable during the summer months, while the Mediterranean generally deteriorates in its appeal for the holidaymaker: the temperatures may become too hot, tropical diseases may become prevalent, there may be water shortages, the landscape may become arid, and freak events in the form of flash floods and forest fires may become more frequent. The coast may become eroded and low lying coastal amenities such as resort complexes and golf course inundated. As a result, this mass movement of tourists could gradually slow, with northern Europeans holidaying either domestically or at least increasingly*

within northern Europe. Equally, southern Europeans may travel north to escape uncomfortable summer conditions at home.

Equally important will be the actual response of tourists to these different scenarios although many studies have shown that tourists rate climate and weather lower than some other variables and that they are more tolerant of extremes than often supposed (Lohmann, 2001; Lohmann & Kaim, 1999).

As climate is a global system, any change is unlikely to be confined to a small number of zones and so deterioration (or amelioration) of the climate and weather in the usual reception region will need to be balanced against changes in the tourist's home country. Any calculus for the effect of tourism on climate change – and, indeed, on many other aspects of environmental impact – has to take into account the fact that while the tourists are away from home they are having much less direct effect on their home environments.

If some of the changes in climate currently being proposed occur, the question then has to be asked: Why travel to the Mediterranean if the north European countries see markedly improving summers (Perry, 2000, 2004)? There may also be technological responses from resorts and hotels to counteract changes in climate in order to sustain demand but these to date have relied on increased energy consumption, itself less sustainable. Following Perry, we may identify drought, heatwaves and sea temperatures as three areas of concern for Mallorca. The first could occur, as it did in the early 1990s, with reductions in rainfall and increases in average temperatures, leading to increased rates of evaporation. Some of the effects of water shortages were described earlier. The second and third areas may make beach-based holidays less attractive – or even impossible – in the high season by 2050, but this effect could be offset by more attractive conditions in the 'shoulder' seasons. For this to happen, north European holiday and work/school timetables would have to change, a serious cultural movement (Perry, 2000).

Rising sea levels for the Mediterranean caused by global warming's effect on rates of ice cap melt and sea water expansion are proving difficult to model; salinity and relative land level changes also have to be taken into account. Estimates vary according to the change agent, but Marcos and Tsimplis (2008) foresee a possible range of +3 cm to +61 cm caused by temperature changes. Most analyses show considerable variation across the Mediterranean. For Mallorca, the vulnerable areas before the end of this century are the low-lying *marinas* to the north and south of the island and the *cales* and coves of the east and south-west coasts. A rise of half a metre would obliterate nearly all of the island's current beaches.

It is disturbing to note that while the Llibre Blanc of 2009 has much to say about pollution of land, sea and air, it has few references to climate change in its final part 'Challenges for the future' (Llibre Blanc, 2009: 365).

Eco Taxes

Besides technical solutions another attractive method of combating the causes of environmental decline is taxation whereby prices are increased to deter poor environmental performance or as a source of revenue to mitigate their effects. More than 40 different types of taxation of tourists have been identified by WTO, many of them associated with entry or exit (WTO, 1998). The industry has generally been opposed to them, but tourist regions see them as a means of offsetting environmental costs. Mallorca has attempted to use fiscal measures to address the many environmental problems that mass tourism has brought in its wake. The best known was the ill-fated ecotax. Introduced in April 2001 it was essentially a room tax, with tourists or tour operators paying a fixed amount based on length of stay. Surpluses were to pay for environmental improvements and to go towards improving goods and services. In effect, the rate levied was too low (one euro per tourist per day) to have any real environmental benefits. Calculations by Aguiló *et al.* (2005) showed that while such a tax had the potential to yield a 10% increase in the Balearic government's income, it might have led to a potential reduction of more than 117,000 overseas tourists. It was opposed by many sectors of the industry, with tour operators threatening to go elsewhere in the Mediterranean. The ecotax was repealed in 2003 (Palmer–Tous & Riera, 2003). Theoretically, it remains a sound idea, but in the case of Mallorca and the Balearic Islands perhaps it was another example of what Picornell and Picornell (2005: 110) have called 'la manca d'enteniment entre la classe empresarial turistica i el Govern del Pacte de Progres' ('The tension between the tourism business and the Balearic government under the Pacte de Progres', the left of centre coalition). It grew out of 'un gran polemica, sobretot per la gran discrepància del sector hoteller, i s'ha convertit en un llei emblematica' ('a great controversy, above all the disagreement of the hotel sector having to implement such a symbolic law').

Others have suggested imposing taxes on other aspects of the tourism industry, including taxing car hire and varying VAT rates. A daily tax rate of 4.0–5.0 euros on hire cars based on engine size and/or distance travelled could both discourage car use by tourists and raise income, depending on elasticity. VAT on hotel rooms in Spain is only 7%, compared with the general rate of 16% in 2008, that is, 44% of the general rate. (Note, that during the financial crisis of 2008 these rates were revised upwards, see p. 143) If areas such as Mallorca had the power to vary this rate on hotel rooms to, say, 12%, it might prove beneficial; anything higher might prove detrimental (Gago *et al.*, 2009: 381–392).

Recently, attempts have been made to impose some kind of environmental tax on aircraft movements, but, as was pointed out earlier, these largely originate in areas other than Mallorca who are sending tourists to the island. Germany is becoming particularly active in this field, but it is a policy being strongly resisted by the airlines.

Fiscal measures, then, appear to have had limited success. Perhaps, the move towards a more diversified tourism product with a lower density of operation, less seasonality and the use of resources different to 'sea and sand' will have a greater effect in countering some of the deleterious effects of tourism. (see Chapter 8). Meanwhile, this chapter has tried to show that the Mallorcan authorities, and to a large degree the general public, have put the concept of sustainability at the centre of their critique of their principal economic activity. While there is now a clear understanding that many of the disbenefits of tourism already present in terms of their environmental effects, there seems to be few real attempts by local political authorities to put controls in place that are aimed primarily at tourists per se and some of the tourist infrastructure, particularly the smaller hotels, bars and cafes. Mallorcans themselves may be very conscious of their own environmental effects, but tourists remain largely unaware or unconcerned. They give little attention to their mode of travel to, from and within the island, their waste production, water consumption and their impact on imports. Perhaps in the environmental sphere, there is a dichotomy at work and in practice it is not possible to have selective policies and legislation aimed at sojourners alone. Discrimination of tourists through taxation, for example, appears impracticable, but raising their consciousness about their environmental impact through education and publicity could have some effect. A better strategy might be for the kind of appreciation and action that may exist in the home country regarding concern for the environment to be applied wherever the citizen resides. The four 'Rs' of environmental concern should be packed in the suitcase alongside the suncream.

Chapter 7
Policy and Planning for Tourism

The Origins of Planning Concepts in Mallorca

The onset of mass tourism in Mallorca in the late 1950s and early 1960s was described earlier as having a transformational effect. It created new forms of employment on a large scale for both locals and immigrants that affected the language and culture of the island. It exploited natural resources, especially land and water, and made new and very substantial demands on energy production. For an island that had always had difficulty feeding itself, it stimulated the massive importation of food-stuffs. Above all, of course, it transformed the landscape of large parts of Mallorca, creating new urban forms and functions and generating a new network of communications. However, at its heart, tourism is a private sector, market-based activity and while it may be true that in its entirety this transformational activity may not all have been planned, large parts of it were and have continued to be. In this sense, Mallorca has been part of the growing participation of the state in the affairs of its people that has its origins in 19th-century concern for social and economic well-being.

The term planning, rather like the argument used earlier for the examination of sustainability (see Chapter 6), cannot be seen as being applied to tourism alone. While tourism, in the mass consumption sense, may have been all-pervasive in Mallorca, its aggregate and external effects have changed practically all aspects of life in the island, most of which have been subject to 'planning' to a certain degree. There is a temptation to confine attention to what the British call town and country planning, which has essentially been land-use planning, and while this aspect of planning will certainly be examined in more detail, perhaps the French term *'aménagement du territoire'* might prove closer to what Mallorcan society, through various government agencies, has demanded and experienced. The history of planning in France has been concerned with the attempt to reduce spatial inequalities while encouraging economic development. In many ways, the Mallorcan experience of what is termed *l'ordinació territorial* is closely allied to French concepts but at the same time has dealt with land-use issues. Historically, these approaches to planning came relatively late to the Balearic Islands. This is partly because for much of the period before the mid-1970s Spain was under governments that had little concern for the welfare aspects of land use whose origins lie more with socialist ideologies such as those that pervaded post-war France and Great Britain. The process of urbanisation associated with tourism came relatively late in the island's history. In addition, there was no intermediate territorial level of planning that could set a spatial context for land use. Lastly, the paucity of national legislation meant that much of the decision making was by state fiat. In fact, and in practice, this was carried out at the level of the municipality (see map of municipalities at the beginning of the book), an ancient level of local government but which in the case of Mallorca had poorly qualified technical officials and political representatives who were subject to many non-political pressures. In theoretical terms, then, until the coming of autonomia and a democratic form of national government from the late 1970s, Mallorca was characterised by procedural and liberal economy planning theory, the first too rigid to be effective and easily subverted, the second antithetical to planning ideas since it was wedded to the benefits of untrammelled market forces. It was not really until the late 1980s and 1990s that an institutional system became more pronounced following left-of-centre coalitions in the *Parlement* and the evolution of a new technocracy well educated and well versed in planning principles and practice. But, by then, of course, as will be appreciated, the excesses of the eruption of tourism in the 30 years from 1955 had taken place and transformed the Mallorcan landscape in so many deleterious environmental ways. The evolving institutional approach has given planners a better-defined role as mediators between market forces, political ideologies and social and economic welfare objectives (Govern Balear, 1997: chap. 13; Rydin, 1993). Over the last 20 years, planning in Mallorca has begun to benefit from legislation and practice similar to that in other west European countries. Most planning and related policy activity is devolved to autonomous communities such as the Balearic Islands' *Govern* and to a lesser extent the island *Consells*. There now exists an intermediate island level of territorial plans, specialised planning policies relating to tourism and much stronger development control practices at the improved municipal level. None-theless, to an outside observer the hierarchy of governments from the European Union through national, autonomias, island councils and their sub-regional bodies (*comarca*) to municipalities seems excessive, making planning more difficult to implement.

   Territorial and land-use planning for tourist environments has to be different because what is being planned is, in many ways, unlike traditional activities. A large part of the urban structure is made up of hotels and apartments, which make different demands on the planning process from the more usual houses, offices, factories and public open spaces of the 'traditional' town or city. Physically, they are different structures that have to fulfil different social and economic functions and mostly for only very limited parts of the year. Land-use values are therefore different. How do non-touristic land uses for local, permanent residents accommodate themselves within the more dominant environment? Rural touristic spaces are by definition certainly not solely agricultural and often barely rural in character but the more traditional land uses remain. Picornell and Picornell have noted for Mallorca that these 'holiday spaces' are also business spaces but 'un espai turistic no es un espai urba normal' ('tourist space is not normal urban space') in which the tourists are temporary migrants; the 'leisure zones' of tourists conflict with the 'service zones' of residents – each makes different demands: 'Hoteles, primeres i segones residences, locals commercials, disoteques... com a parasits del turisme, sense ordre ni control i de manera massiva i intensiva' ('hotels, first and second homes, shops, discotheques ... are the parasites of tourism without order or control and massive and intense in style') (Picornell & Picornell, 2002a,b: 45). Is planning one way to resolve or at least mediate between these conflicting interests?

Planning and Political Economy

While other land uses and economic developments other than tourism clearly occur in Mallorca, the dominance of that one activity has shaped most aspects of planning until recently. The planning process has been in part a late response to market excesses and has become a means of developing more socially and politically acceptable patterns in the future. As in nearly all societies, this aspect in terms of enacted legislation and its translation into practice is subject to political ideologies. In Mallorca it has been reactive and proactive in its expression, often confusingly, at the same time.

In addition to planning land uses and spatial distribution, there are also some mechanisms whereby Spanish and Balearic Island governments have tried to manage aspects of the tourism industry for reasons of political economy. Four might be identified: the need to increase the supply of foreign exchange, originally to boost the fortunes of the peseta; to improve Spain's standing in international relations within Europe and a wider world; to provide an improving infrastructural base upon which private enterprise could flourish and finally in the Balearic Islands themselves to manage better their dominant economic activity, maximising income while minimising social and environmental costs. While these may not be suggestive of 'planning' in either the territorial or command economy senses, these show how the state had become involved in tourism as it moved from a centralist position to one of democratic devolution. In this sense, policymaking might be a more appropriate term.

'Planning' before Planning

There is a long history in Mallorca of the imposition of order on its landscape and its people. Mallorca, like so many islands, is the product of invasion and settlement, of the inward movement of ideas and their adoption and adaptation to island circumstances. From the patterns of Talayotic settlement in the late Bronze Age, through the Roman imprint and later Arab and Berber settlement in rafals and alqueries through to the planned new towns established by Jaume II in 1300 AD, there have been clear attempts to produce order in the landscape. The reform of agriculture in the 18th and 19th centuries and the growth of the new estates centred on the *possessions* produced much of the rural landscape that we see today. With industrialisation and the growth of Palma in the late 19th century came the more familiar problems of industrial location, pollution and unsanitary conditions, and town planning. The founding of the so-called garden city resorts in the 1920s and 1930s exemplified the design and layout approach to planning (Seguí Aznar, 2001). The growth of tourism, then, from that period and its explosion after the mid-1950s was in many ways continuing a long history of landscape transformation, the result of a new form of tertiary activity. Mallorca has been something of a pioneer in trying to reconcile the considerable social and economic benefits of tourism with the environmental degradation that it has caused. This involved trying to manage the economic activity itself as well as its locational and environmental effects.

Planning for tourism in the era before the 1940s in Mallorca was largely in the hands of the resort developers described in Chapter 3 who acquired land and real estate from the declining nobility and aristocracy and built holiday complexes and villages following British and American theories. Earlier, hotel building in the period to about 1930 was largely speculative either on land near attractive physical resources with good access to Palma or in the city itself where the authorities usually took an encouraging and tolerant view as they began to appreciate the economic benefits of this new tertiary activity. What control there was was in the hands of the municipalities, historically often potentially powerful agents, but their interests were extraordinarily local, even competitive amongst themselves, and often subject to corruption. Above all, as the first 'boom' was to prove, they simply did not have the technical expertise in fields such as planning.

As in most west European countries, the subject of planning as defined above really developed in conjunction with industrialisation, urbanisation and population growth and tried to deal with their environmental impact. On the one hand, the objectives of such planning were to provide a more efficient environment in which business and commerce could flourish and on the other to give better living conditions for society's general well-being. In both cases, planning was the instrument used to distribute resources in ways that market forces and the price mechanism apparently could not. In Mallorca's case, industrialisation was a relatively late introduction and the factory system in many ways still had not had much effect on production. The urbanisation associated with factory working was on a small scale and was largely confined to the textile industries and boot and shoemaking, themselves geographically localised in Palma and to a much lesser extent in smaller towns such as Manacor, Llucmajor and Inca. However, the period from the late 19th century through to the early 1930s saw a considerable growth in Palma's population, the expansion of the transport infrastructures, particularly railways, road and the port, and an increase in the effects of the tourist trade that dates from the 1900s. Modern, factory-based industrial development in Palma was originally located in areas close to the old walls but was later sited in the new *ensanche* and in new suburbs such as Santa Catalina and Soledat (Carbonero Gamundi, 1991: 98). The confinement of most of this to Palma meant that its urban infrastructure came under increasing pressure including a need for more and better housing to accommodate a rising population. The first large-scale plan to attempt to deal with this was Bernat Calvet's putative *ensanche*, a suburban expansion made possible by the demolition of most of the city's walls in 1901. This solved only some of the problems of sanitation, housing and industrialisation; he appears to have been much more concerned with improving the radial routes that became focused on the medieval city centre leading to intolerable congestion (Seguí Aznar, 1999: 57). His was a plan on paper more than one that led to actual building (Ruiz-Viñals, 2000: 95ff). It was not until the adoption of the plans drawn up by Gabriel Alomar in the 1940s by the city authorities that substantial progress along modern planning lines was made.

By this time the political climate for planning had changed dramatically with the outcome of the Spanish Civil War (1936–1939). As under Primero de Rivera in the 1920s, Spain had once more reverted to a strong oppressive state under the fascist regime of Francisco Franco. Much of Europe had also come under the spell of centralism whether communist as in the Soviet Union or fascist as in Italy, Germany and Spain. In all these regimes, planning was held to be a central concern of government. This was especially true in post–Civil War Spain because the Madrid government saw the reason for the 'failure' of the Republican government (1931–1936) and the need for armed intervention coming from demands for autonomy and devolution in Catalunya and the Basque country. Whereas the Soviet and Nazi regimes might be characterised by their ruthless efficiency, Franco's fascist era was chaotic, particularly with regard to planning, being so dependent from the very beginning on captains of industry and business as much as of the military variety. Many of its objectives for the state's development were subjected to growing capitalist pressures. In town planning and architecture, this meant permanence, massiveness and authority. A good example in Palma was the new major commercial street of Jaume III (originally Jaime III as the Catalan language was banned under Franco), together with an emphasis on road building to permit more efficient distribution services for commerce and industry.

However, it became increasingly clear to some that tourism as an economic engine could perhaps prove to be additional to industrialisation for Spanish economic progress. From a quite early date, although to a limited degree, tourism became an instrument of national economic policy. Until the coming of autonomia in 1983, in the case of the Balearic Islands tourism policy and tourism planning became closely allied. There was a strong centralist tendency within which local – Mallorcan – aspirations had to operate. This was always modulated by the local interests and as Cirer has shown Mallorcan entrepreneurs felt that their aspirations were in advance of those of the state. In hotel building, the improvement in marine communications, promotion and in the founding of the Foment de Turisme Mallorca had shown innovation and investment that led to early advances in tourism development (Cirer, 2009: 203–232). Pack has demonstrated the importance of a national context. From the early part of the last century, Spain was beginning to appreciate the considerable economic benefits that could be derived from tourism through the establishment of the National Commission in 1911 and the Patronato Nacional de Turismo in 1928, which sought to join business interests and the state. Some limited subsidy to tourism became available, but the era's greatest legacy was the network of paradores from which the Balearic Islands did not benefit (Pack, 2006a: 27–30). Under the Republic (1931–1936), the national economy as far as tourism was concerned was affected by the downturn in world trade following the Great Crash of 1929, reducing the inflow of international visitors. However, many saw tourism at this time as appealing to a more popular and national market, a sector that was perhaps less prone to deflationary pressures. The short-lived Republic itself did little to foster such tourism, but Mallorcan businesses worked together to develop a substantial share of the national market (Buades, 2004: 57). Following the fascist success in the Civil War, this view was not shared initially by the hierarchy who saw modernisation of the state being achieved via industrial development not holidaymakers. The ultra-conservative cultural ideals being inculcated by the regime, supported by the Catholic Church, would also, it was thought, be undermined by an influx of north Europeans. In the 1940s, the Spanish economy was dedicated to self-sufficiency and autarky – or in essence, isolation, because many believed that an open economy would undermine Falangist values. A second issue for the Franco regime was its political isolation, although it had been ostensibly neutral during the Second World War. Largely thanks to Allied pressures, it remained in the 1940s something of a pariah. Many in central government saw tourism as a medium for improving international relations. Eventually Franco – reluctantly – began to see that the economic and political benefits of developing tourism would outweigh the social costs (Pack, 2006a: 41).

Mallorca because of its success with tourism before both wars was in an ideal place to benefit from these changes in government policy. Madrid's role was essentially one of improving the infrastructure for tourism. Since most tourists reached Palma by road and ferry, these were two areas of state investment that benefited. As shown in Chapter 3, there was also a need to improve the supply and appearance of hotels. An improved classification system was introduced with an emphasis on improving middle-ranking hotels and pensions, but many hoteliers believed it too rigid in its demands, preventing a more rapid response to growing tourist requirements. However, as previous chapters have demonstrated, the state under Franco was obsessed with the need to build foreign exchange, leading to the 1959 Stabilisation Plan and the devaluation of the peseta. Both proved vital for the burgeoning tourism industry, the latter establishing a truer value of Spain's currency, which quickly made holidays in Spain about 15%–20% cheaper. This encouraged British tourist numbers to Spain to rise by a quarter by 1960, increasing the country's Sterling reserves by 250% (Pack, 2006a: 88).

Planning in the Modern Era

In order to improve the national planning of tourism, it was necessary for government to introduce a new technocratic bureaucracy to understand the private sector processes of resort development, to collect data on numbers and financial flows, to encourage inward investment from overseas and to raise standards. It also meant establishing a proper political identity for tourism within government. Until 1962, tourism had been something of an unwanted child, being moved from ministry to ministry, but with the appointment of Manuel Fraga Iribarne (at the young age of 39) to the newly formed Ministry of Information and Tourism, the industry was placed on a firmer footing (Mata Pastor, 2002: 82). With the state now intent upon modernisation, the emphasis for its role was placed on infrastructure, to be achieved via the new Social and Economic Development Plan of 1964–1967, which would benefit many aspects of Spanish economic development besides tourism. It focused state investment on roads, railways, water supply and irrigation and electricity supply. However, Fraga's legislation, which aimed at trying to protect the municipalities from poor-quality physical development, proved too weak and it was often opposed by other government departments. While some developers were prepared to undertake road building to join an inland municipality to its new coastal resort, most strategic roads linking the coast in Mallorca to the capital were not built until the 1950s and 1960s. Planning was more noticeable by its absence; most developments were approved only retrospectively. Many municipalities had little or no means of controlling resort developments, and developers were ruthless in their illegality and rapaciousness – in defiance of national laws: 'The mid 1960s witnessed the zenith of lawlessness' (Pack, 2006a: 173). The era of the first boom until 1973 has been described by Rullan as 'un convulsió territorial' in which the urbanised area more than doubled to 120 km², a new airport was built and immediately expanded, new water supply facilities were constructed and electricity generating capacity massively increased. In effect, it was only these large capital investments that could be controlled by planning; resort development remained largely anarchic. There was little democratic control over development. In principle, planning applications had to be approved by the Provincial Planning Board of the Balearic Islands, but this was purely indicative since most local authorities had no municipal plans within which to set such applications. Mallorca's 63 municipal authorities did not seriously become involved in the planning process until 1975 or later (Rullan Salamanca, 2007a: 20–23).

Following the effects of the 1973 energy crisis, the tourism industry in Mallorca expanded more slowly; 257 establishments were actually closed in the period 1973–1978. The 1980s then saw an accelerating second boom period, leading to more and different planning pressures. If the 1960s and 1970s were characterised by hotel building, this next phase witnessed a real-estate boom. Chapter 5 showed that this was the era of apartment building largely to house a new form of tourism but also to accommodate Mallorca's growing resident population. In the Balearic Islands, the number of bedspaces rose by more than a third from 1982 to 1988 but the non-regulated part of this increased by two-thirds. By 1988, the percentage of non-regulated places rose to 14.6% of the islands' total (Picornell & Picornell, 2002b: 97). In addition, this was a period of considerable second-home building, much of it in unspoiled countryside. Once again, planning legislation seemed unable to control this illegal growth. The Cladera Acts of 1984 and 1987 (see Chapter 5) were aimed at controlling hotel development but that was now not where the pressure was coming from. The Special Areas Legislation to protect natural areas (1984 and 1991) and the creation of the first natural parks (Tramuntana and Albufera) were designed to prevent further expansion on the fringes of resorts but were largely ineffectual in preventing villa expansion in the countryside. At the same time, new tourist activities were developing in rural areas such as golf, agrotourism, cycling, bird watching and hiking; the first of these was nearly always accompanied by second-home building, which the legislation seemed unable to regulate. The expectation that the devolution of most planning following autonomia would immediately resolve many of these planning issues and reconcile tourism to the environment was premature. The weaknesses in the planning system during the 1980s are largely ascribable to a number of factors that included the absence of any island-wide territorial plan to act as a constraining context on local development, the speed with which developers could erect apartments and second homes, the lack of a cadre of trained planners and planning expertise at the level of the municipality, the frequent coincidence of political control and the construction industry and the continuing and increasing dependence of Mallorca on the tourism industry.

From the late 1980s, the context for planning began to improve in that the continuous growth of the industry was being questioned not for the first time but now more consistently. Rather than simply managing growth, public pressure suggested better attempts should be made to contain, reduce or restrict it. A significant difficulty was that while tourism might be more severely constrained there was a conflicting demand for other sectors of the space economy to expand – to house the growing population, to develop new industrial zones (poligonos), to relieve road congestion and to improve much of the infrastructure that underpinned the non-touristic economy. By 2000, the urbanised area of Mallorca had already reached about 250 km^2 and a population of more than 700,000. One major policy document that illustrated this shift in emphasis was POOT (Plan d'Ordinació de la Oferta Turística; Govern Balear, 1995). It was perhaps the first real plan that attempted to contain the growth of tourism and to improve its infrastructure on an island scale. It also recommended better planning to protect fragile environments, particularly overcrowded beaches, to limit the density of total population in tourist areas and creating something akin to green belts by protecting the fringes of coastal urbanisations. More radically, it recommended freezing land for new urbanisations and even to begin the demolition of illegal hotel development and converting some older ones to apartments for local people. Some land already labelled for urban expansion in previous plans was declassified including more than 1600 ha in Calvià. POOT began in 1989 but took seven years to be approved (Amer i Fernandez, 2006a: 108, 2006b; Rullan Salamanca, 2007b: 37).

During this long interval, the political representation in Mallorca began to change with a new alliance formed on the Left in the island's Consell to be joined by a more progressive coalition in the Govern from 1999. A number of important acts can be identified from this period: the Lei de sol rustic (1997), the General Tourism Law of 1999 and importantly DOT (Direció Ordinació Territorial) of 1997. Rural land was designed to limit building to farm workers' housing but was often subverted by disguised second-home building, leading to attempts to limit construction in the countryside by imposing minimum plot sizes. Any further building in rural areas was supposed to be in existing settlements, thus following an age-old principle in Mallorca of nucleation in preference to dispersal of settlement (Blázquez Salom, 2006: 165). DOT began with a wide-ranging analysis of growth and development up to that date but with a diagnosis based on planning within the principles of sustainability that were examined in Chapter 6 and, by then, gaining widespread influence following Bruntland (Govern Balear, 1997: 14–21). Principally a constraining plan, the General Law of 1999 that followed it laid down that there were to be no new urbanisations, any development of land classified as urban was to be at least 500 m from the coast, no municipal authority's urban land was to be expanded by more than 10% and the classification of almost 3000 ha of land for urban use was to be suspended (Seguí Pons & Martinez Reynés, 2002: 274). Following DOT and the General Law, political tension between the state or national government increased, whose liberal economy was being pursued by the conservative Parti Popular; Balearic and Mallorcan governments were under the progressive PSOE/UM alliance. Constraining the growth of tourism per se may have been the objective of much of this planning, but it often failed to recognise the rising demand for 'quality-of-life' improvements for local residents, such as better sewage and waste disposal facilities, improved water and energy supplies and above all, more and better housing under rising population pressures. Mallorca's population rose from 614,000 in 1990 to 760,000 in 2004. In addition, there was also a growing demand for spaces for the new forms of tourism beyond the mass model such as for golf courses, water parks, aquaria, and marinas. The construction industry was, and remains, a very powerful lobby (Picornell & Picornell, 2002a: 42, 2002b), and construction had become a very important part of the labour market, with its employment rising from 29,000 in 1995 to 57,000 in 2001 (Rullan Salamanca, 2005).

In this century, planning has again shown its vulnerability to political ideologies and policies. In the field of transport, coalitions of the right have favoured road building programmes, which have included the extension of the airport motorway to beyond Llucmajor, the dualling of the main highway between Palma and Manacor, the completion of a network of high-speed roads around Palma northwards beyond the Via Cintura of the 1970s and the expansion of many cross-country roads linking towns in the *part forana*. Coalitions of the left, on the other hand, have invested in public transport: the expansion of the railway to Sa Pobla and Manacor, the creation of the underground Intermodal Exchange Station next to Plaça d'Espagna in Palma, the creation of TIB as an intergrated long-distance bus service and the expansion and improvement of Palma's public bus network linking expanding suburbs and south coast resorts to Ciutat. In nearly all these, there seems to have been a certain grandiloquence while many of the smaller, more local planning issues concerned with resort development and second homes in the countryside have been neglected. Few of these grander schemes appear to have been orientated towards the tourism environment.

Planning, then, has not been the panacea to resolve Mallorca's environmental ailments. At the larger scale, tourism is a business activity crossing international boundaries and in Mallorca it has been so dominant that management and regulation by the state rather than the market has proved difficult. At the more local scale and in terms of land use, planning legislation came too late and was almost impossible to apply in order to control the excesses of the 1960s. The landscape of tourism today may have a better management framework, but it is still subject to the vagaries of political ideologies. As the planning emphasis shifts more towards environmental sustainability, serious questions are now being asked about the scale of tourism on the island, the very characteristic that defined its mass appeal.

Chapter 8
Economy, Business and Politics

Introduction: Some Macroeconomic Aspects

As Mallorca's economy is so dominated by the tourism industry, it is hardly surprising that the majority of businesses on the island are in that sector or closely allied to it. The three main elements are hotels, restaurants and bars, but intimately connected are self-catering accommodation such as apartments, transport systems, clubs and discotheques, certain retailing activities, outdoor entertainment and many recreational leisure activities. Of increasing importance are sailing and water-based activities. This might be summarised as TAFL: transport, accommodation, food and drink and leisure. Such a litany is familiar to all tourist areas, but in Mallorca it is the scale of each and the totality of the whole that is so dominant. For example, in 2008, in Mallorca there were 1587 establishments offering some kind of accommodation containing 285,370 places, 2927 restaurants seating 202,124 diners and 1588 cafes and 3558 bars (*Dades Informatives*, 2008: Tables 7.7, 8.1, 8.3 and 8.5). Most other parts of the economy are also linked to tourism though less directly; the most significant are the construction industry and banking. With the exception of construction, all of these parts of the economy can be described as either tertiary or quaternary. But even the primary and secondary (manufacturing) sectors are dependent in part upon the tourism industry; for example, horticulture within agriculture provides a limited proportion of fresh foodstuffs for hotels, apartments and shops and the food and drink sectors of manufacturing provide a similar supply of goods. A longer list is possible because so many economic and business activities are either forward or backward linked to tourism. But, of course, the island economy is by no means self-contained and is far from self-sufficient. Any description or analysis must take account of imports, after all the main source of 'national' income is in fact an import – the tourist from overseas. With a region whose resident population increases during the year from less than 1 million to a total visitor population of nearly 9 million with major concentrations in mid-summer, there is little chance of such a small resident population and workforce being able to supply so many goods and services from its own resource/demographic base. It is dependent upon imports for most goods and temporary seasonal labour for many services. *Valor afegit brut* (VAB), that is, value added during production (which is roughly equivalent to gross domestic product for tourism), in the Balearic Islands is 41%, down slightly from 1983 (43.2%), of which 26.6% is attributable to non-residents (primarily tourists) (Llibre

131

Table 8.1 Islands services and tourism

Services related to passenger transport	90%
Renting/leasing property	90%
Vehicle hire	70%
Renting of second homes	70%
Air or transport passengers	50%
Restaurant services	42%
Alcoholic drinks and tobacco	27%
whereas many basic industries indigenous to the islands are much lower, viz.	
Textile production	22%
Agriculture and fishing	21%
Leather and shoes	16%
Glass and ceramics	7%
Cement and plaster	4%

Source: Data for 2004, Llibre Blanc (2009: 40)

Blanc, 2009: 39). Table 8.1 gives some idea of the dependence of the islands' economy on non-residents as a source of value added in terms of services.

At the macroeconomic level, the Balearic Islands, and especially Mallorca, have experienced growth well above the national average for the last 20–25 years. Between 1987 and 1995, the annual average growth rate was 3.7% compared with Spain's growth rate of 2.8%. In the 5 years to 2000, national economic performance improved but the Islands still had a growth rate 23.2% above the nation's. By 2006, the gap between them closed markedly, but the Balearics still grew more than 10% above the national average, placing them in fifth position within the autonomous regions of Spain. While this has all been achieved primarily via the tourism industry, there has been a continuing process of development in the tertiary sector overall. It commands a remarkable 81% of the regional gross domestic product compared with only 67% in Spain as a whole. Because of the ways in which tourism and the tertiary sector have operated in the last quarter century, output in agriculture and manufacturing have declined relatively speaking. In the construction sector, the Balearic Islands and Spain have almost doubled their output but considerably above the current level of demand, leading in part to its collapse after 2008. The success of tourism and the rest of the tertiary sector, the dependence of most economic activities on it and the lack of significant exports have exposed Mallorca to the criticism of extreme economic monoculturalism. However, the economy is becoming less specialised and more diversified. In 1986–1990, the index of specialisation for the Balearic Islands against the national structure was 0.224, but in 1995–2002, it was 0.212; the coefficient of diversification fell from 0.712 to 0.661 over the same time period (Llibre Blanc, 2009: Table 3.1, 134).

Clearly, for such success Mallorca's tourism industry must have been very efficient in terms of value of each unit of output – costs of production have been kept relatively low and returns on capital invested relatively high. However, the returns in the last decade have been slowing down, a worrying sign for such a dependent economy. This, combined with the downturn in the world economy from 2007, which has especially affected likely future spending on leisure generally, prompted the Balearic authorities to commission a detailed enquiry into likely trends in the future of their tourism industry – the second Llibre Blanc (2009). The rush into construction as an alternative initially appeared sensible given the rate of population growth identified previously. This generated an increased demand for labour that was satisfied only by further immigration shortly after which (2008–2009) the sector began a rapid decline. Housing completions had risen from about 7000 units in 2002 to just under 12,000 at the peak of the boom in 2007. In 2008, this fell back to less than 5000 (Social and Economic Report of Balearic Islands, 2008: graph 7.1, 129). Table 8.2 shows the impact of the recession from 2007 to 2008.

The decline in nearly all sectors, but especially in construction, was in part compensated by a relatively small decline in tourist numbers. In fact, the number of German visitors rose by more than 5% in 2007–2008.

The need to diversify the economy may be obvious but given such dependencies, into what? The kinds of innovation in recent years, as we shall show in the next chapter, have largely led to further additions to the leisure industry – golf, sailing, conferences etc. – that are, in effect, 'more of the same' and so far the income and employment generated still pale into insignificance when compared with the staple of mass tourism. Having been so dependent upon 'the visitors' for so long, there is little evidence of investment in real alternatives to tourism and whatever there is is mostly in the public sector such as the new Metro, new hospitals, the expansion of the university and road building programmes. The most rapidly growing area of employment is the public sector thanks to Spain's and the Balearic Island's very bureaucratic system of government at all levels: national, autonomous region, island and municipality – understandable in a period of recession perhaps but the system predates the current downturn by nearly 40 years. In the current era of austerity (2010–2011), it is highly likely that the size of this public sector activity

Table 8.2 The recession 2007–2008: Percentage changes in selected sectors in Balearic Islands

Sector	2008	2007
Petroleum-based consumer products	− 7.1	− 6.7
Petrol	4.8	− 2.4
Aviation fuel	− 5.0	8.6
Diesel	− 2.7	5.7
Vehicle registrations	− 34.5	3.1
Housing starts	− 62.2	0.6
Flights	− 2.0	3.7
Freight flights	− 6.4	1.7
Passenger flights	− 1.7	3.7
Tourist arrivals by air	0.9	3.4
British	− 1.1	3.2
German	5.0	5.3
Nights in hotels	− 2.0	− 2.9
Seaborne freight	6.7	2.5
Passengers by sea	5.3	10.1
Ferries	− 16.8	7.4
Cruises	7.9	13.5

Source: Social and Economic Report of Balearic Islands for 2008 XLI (September 2009)

will have to be reduced. It is also the case that non-residents increase the need for spending by local government but without the full complement of the necessary tax receipts. A previous generation of investment in, for example, ParcBIT – a high-technology business park – seems to have produced little by way of new product industries, and its impact on information technology appears to have been mainly directed towards tourism and its adjuncts. Investment from overseas is similarly limited to activities related to tourism and consumption by non-residents, although in the retail sector consumption by Mallorcans has increased as their affluence has risen on the back of tourism. German and French investment in retailing is also to be noted.

Finally, Mallorca's tourism industry has always been sensitive to currency fluctuations. In the past, the relative pricing of the guilder, the franc, the deutschmark and the pound affected negotiations between tour operators and local hoteliers. With the coming of the Eurozone in 1999, many of these difficulties were lessened even though the lack of price parity between different European countries and Spain affected spending levels. In the past, the pricing of certain goods in Mallorca relative to prices in the countries of tourists' origin always added an economic attraction to holidays there; this was especially true of tobacco, spirits, petrol and latterly, property. The adoption of the euro, however, appears to have led to increasing convergence in the pricing of those goods sensitive to non-resident demand. The important exception to this convergence for Mallorca has been UK's retention of sterling, which was not a problem as long as its relative position against the euro remained positive. However, the fall in the value of the pound against the euro since 2008 by at least 20% has militated against the island's second most important customer. It may prove difficult for Mallorcan hoteliers to negotiate appropriate prices against such a fall.

Employment

Naturally, the employment structure of Mallorca and the archipelago is similarly dominated by tourism and the tertiary and quaternary sectors. Two characteristics are to be noted at the beginning. One is that much of the work in the tourism industry is relatively poorly paid compared with other activities and the second is that much of the work is seasonal. A third observation is that the small population of the island, historically speaking, has been an insufficient labour pool for the tourism and the construction industry. Hotels, restaurants and bars have led to temporary, and increasingly permanent, labour immigration. As hoteliers and barkeepers have kept up pressure on reducing labour costs in order to remain competitive, many local workers in tourism have gradually 'migrated' out of the low-skill jobs in hotels and restaurants into often better paid and more conducive employment in offices, banks and all levels of government, perhaps the result of rising education and training standards in Mallorca's schools and colleges. Trade union membership and organisation in Spain may be high, but mobilising a temporary and immigrant workforce in the tourist sector has not proved easy.

Within the three broad sectors of the Balearic Island's economy, the primary sector has fallen from having 9.5% of the labour force in 1987 to 2.3% in 2006. Manufacturing has fallen from 14.9% to 7.1% in the same period, but the tertiary sector now contains 76.6% of the workforce, rising from 66.1% in 1987. Mallorca's own data show slightly more extreme shifts. In part, these moves are a reflection of changing structures in the national patterns of employment where Spain's labour market contained 44.6% in services in 1987 but by 2006 had increased to 60%. Such tertiarisation is a characteristic of all developing economies but, as in so many things, it is exaggerated in Mallorca.

The tourism industry in Mallorca directly employs about 104,000 workers, some one third of the labour force. However, this figure varies considerably on a seasonal basis and so in 2007 the middle two quarters of the year saw 127,000 and 125,000 employed in tourism with the low-season quarters employing 90,000 and 75,000. Even across all sectors of employment, there is a 40,000–50,000 difference between high and low seasons (IBAE, 2007: Employment tables). These data refer to workers registered with the national insurance system, but in Mallorca, as in so many tourist regions, there is a considerable number working in the 'black' economy. Of the legal workforce related to tourism, about 44% is engaged in the accommodation sector, 17% in restaurants, 15% in passenger transport, 11% in bars and cafeterias and 6% in cultural and recreational services. There are also about 15,000 self-employed persons in tourism.

The Hotel Business, the Economy and Politics

In Chapters 2 and 3, we showed that as a business activity the hotel industry in Mallorca had slow and uncertain beginnings and concentrated on attracting the upper and middle classes. It is important to appreciate the historical development of tourism as a business and the emergence of a new entrepreneurial class to develop it (Russell, 2006).

Until the early 20th century, there were few hotels worth the description; it was not until the opening of the Grand Hotel in 1903 that this segment of the tourism industry began to develop along lines well established in other European countries such as France, Italy and Switzerland at least 50 years earlier. Cirer has called this period 'the frontier between two epochs', between the spartan regime offered to early visitors and the more appropriate accommodation offered once the decisions had been made by the Mallorcan business community to embark on a policy for the development of tourism (Cirer, 2009:149). Despite the number of visitors increasing from the 1880s, most complained bitterly about the quality of the accommodation in the island's *fondas* and *hostals*. Miguel Del Sants Oliver was not the only reformist Mallorcan businessman to write of the need for proper hotel accommodation for the type of visitors the island was trying to attract (Cirer, 2009: 142–147). The Grand Hotel was an iconic and symbolic building not only architecturally – its style was in the latest fashion – but its opening ceremony was attended by the great and the good of Mallorcan society, signalling that this kind of commerce was socially acceptable. It marked the beginning of a new era for Mallorca's affluent visitors, but equally importantly it showed Mallorcan business-men that investment in hotel development could be as attractive as in the more traditional industries of textiles, footwear and food processing. As more high-class hotels were built in the next decade – such as the Gran Hotel Marina in Sóller, the Hotel Reina Victoria in El Terreno and the Alhambra in Palma itself – new businesses began to emerge, which were to form the backbone of a new class of entrepreneurs soon to be joined by new activities to service the hotels and their clients.

However, the First World War was to expose the fragility of this kind of development aimed at Europe's wealthier classes. The number of passengers disembarking at Palma – a category that includes not just tourists – fell from 3352 in 1912 to just 102 in 1917 (Cirer, 2009: 187). The social and economic changes that emerged from the new order of the 1920s suggested that a much broader base to this nascent industry had to be found; the grand luxe had to give way to a more democratic form of tourism in which numbers needed to increase in order to command economies of scale. This new clientele would need to have access to a much wider range of facilities and services. The question for the Mallorcan economy was whether or not it could rise to this new challenge. Would sufficient local entrepreneurs and capital be generated by the island to set it on its way to this broader concept of tourism or would it depend on inward investment from the Peninsula and overseas? Cirer may believe that Mallorca invented mass tourism in this next era, but it was of considerable less intensity and scale than that which was to emerge in the 1950s and 1960s (Cirer, 2009: 199). The hotel and tourism business of the years to 1936 was certainly growing as demonstrated in Chapter 3, but to this writer it was a long way from what might be described as 'mass', even if by the mid-1930s Mallorca was the leading destination for tourists to Spain. If anything, 'mass tourism' was invented by British travel agents taking advantage of the needs of the Franco regime in the 1940s and 1950s and it was in this period that the modern hotel industry was founded with its close links to the general political economy of Mallorca and the wider state.

When the Foment de Turisme was founded in 1905, its main aim was to promote the island. There were few, if any, hoteliers on its committee, the majority being journalists, newspaper publishers, men from the professions and general businessmen. During the next 20 years, the membership was widened to include hotel owners and managers such as Adan Diehl of Hotel Formentor, Juan Benesabene of the Alhambra and Jaume Ensenat of the Hotel Victoria, indicating a shift from the general to the specific but still focussed on the large luxurious hotel to the exclusion of the smaller establishments. By the 1930s, the number of tourists staying in hotels rose steadily from 20,168 in 1930 to almost twice that number by 1935. By 1934, in addition to the seven 'grand' hotels in Palma there were 25 other hotels and 46 pensions in the city, and so the de luxe establishments contained only about a quarter of Palma's capacity. The geographical spread away from the south coast was already taking place, with Pollença having eight establishments by that date and Soller five hotels (Cirer, 2009: 273). Although Palma's share of hotels remained fairly constant over the first part of the 1930s at just over 50%, by 1936 there were 62 establishments outside Ciutat. While some of these developments outside Palma were high-class hotels, such as Formentor, many began to cater to less affluent visitors. The spatial spread was accompanied by a more democratic form of supply. Politically, the dictatorship of Primo de Rivera, the monarchy under Alfonso XIII and even under the Republic after 1933 made little difference to the local business environment. As Manera has argued, the commercial middle class of Mallorca was well integrated into European business networks – '... in constant contact with the world of transactions ...' (Manera & Garau-Taberner, 2009: 42) – but they were peripheral to Madrid politics; amongst Mallorcan entrepreneurs, only Juan March held a national position (Amer i Fernandez, 2006a,b: 42). It would appear that there was little interest in Mallorca's tourism industry from Madrid even though tourism in Spain as a whole was emerging as a new economic activity. Perhaps, it was a case of Mallorca forging its own destiny under the influence of its own entrepreneurial class, supplemented by overseas investors – often Latin American – who had strong links with the island. This was to change markedly after the Spanish Civil War and the Second World War.

The smaller and less luxurious hotels may have expanded in the 1930s, but it was the growth of the industry after the Second World War described earlier that really gave opportunity to a new and numerous form of investor. Amer i Fernandez believes that the hegemony of the former bourgeois class of landowners, industrialists, the professional classes in Palma and the remnants of the nobility and aristocracy was added to, and to a large extent undermined by, the emergence of a new hotelier class (Amer i Fernandez, 2006a,b: 46). The traditional, pre-war hotelier ran a small establishment, usually with fewer than 50 beds, conservative and unresponsive to changing demands from the new holidaymakers flooding out of northern Europe in the late 1950s. The Stabilisation Plan of 1959 created new opportunities for a different type of hotelier willing to respond to this new market, willing to take more risks and to provide cheap but good-quality accommodation. In the first 'boom' up to 1973, this new class came from business activities other than hotels or catering, often from the construction industry, for example. Its main difficulty was its lack of capital, which was partially solved by the Crédito Hotelero and by working closely with the tour operators located in the United Kingdom and elsewhere. This more secure financial base stimulated a rash of hotel and resort construction as examined by Seguí Aznar (2001: 17–120) and Buades (2004: 159–160). By 1964, there were more than 900 hotels and pensions in Mallorca. Rullan Salamanca (2005) was perhaps only partially correct in describing such a large number as part of a Fordist system of production in so far as the product was fairly similar; it was more akin to a cottage industry, with each of the small hotels vying for a share of the market generated by British, German and other European travel agents and tour operators who, if anything, articulated something more like a Fordist approach to consumption. After the setback of the early 1970s' energy crisis, it became obvious that cooperation rather than competition would help guarantee the demand side of the equation and lead to, if not a standardisation of the product, then at least an improvement in accommodation standards. This movement was led by the larger hotels such as the Nixe and the Phoenix. Sbert Barceló (2009) has also shown in the Platja de Palma from the outskirts of Palma, eastwards to S'Arenal, that its beginnings were probably on a sub-regional basis within the island. Here, the hotel-owning families worked closely together to promote the resorts in their area as well as their own hotels. Similar movements took place in Calvià and the Bay of Pollença as they developed. From 1970s, all hotels began to organise themselves into federations for mutual benefit in negotiations with tour operators and with trade unions. By 2005, 25 associations made up the Federació Hotelera, which included 80% of the hotels in Mallorca – 934 establishments, 39% of which were part of a chain. Other federations were established for small hotels and for agrotourism (Amer i Fernandez, 2006b: 112). By the early part of this century, even the externality benefits of environmental sustainability had grown up around a common agenda in Calvià and its hotels (Dodds, 2007: 300).

However, in the second 'boom' of the 1970s and 1980s, another movement was afoot in the hotel industry that was more reflective of a Fordist system, namely, the emergence of a number of hotel chains through the process of merger, acquisition and takeover accompanied by their increasing investment overseas (see Table 5.2). By the beginning of 21st century, three types of hotels could be identified: large hotel chains (Sól Melia, Riu, Iberostar etc.), with much of their multinational business overseas; middle-sized hotel chains (Viva, THB, Grupôtel etc.), with an equal share of business in Mallorca and elsewhere and thirdly, small hotel chains and independent hotels, largely island-based (Photo 8.1). Hotel groups such as Hoteles Mallorquines (which later emerged as the Sól-Melia organisation), Barceló and Riu were powerful products of and leaders of this process. However, Sastre has shown that during the second boom, despite this process of concentration, the hotel business remained in many hands. In 1975, 5% of businesses controlled only 38.5% of places; by 1992, the same percentage controlled 53.5% of places and 95% controlled 46.5% (Sastre, 1995: 98).

Photo 8.1 High-rise hotel architecture of 1960s by a major chain in Magaluf

About a quarter of all Spanish hotels are located in the Balearic Islands. Of these, about three quarters are in Mallorca. The current structure of hotels by size, category and geography reveals another important feature: the persistence of the three-star hotel over time despite the supposed benefits of economies of scale. Within this category, the majority continue to have 300–400 beds. The size of the average three-star hotel has fallen marginally from 307 beds to 297 beds, from 2001 to 2008. The larger-sized three-star hotels (400–500 beds) have shifted upwards only in size, with the group increasing from 14.1% in 1975 to 15.8% in 1992 (Sastre, 1995: 108). For Mallorca, by 2001, 55.2% of hotels were three-star, with 60.2% of all hotel places; by 2008, these figures were 52.8% and 58.2%, respectively, showing that in this century this category has lost some ground. The rise of the luxury five-star hotel in Mallorca has had much publicity, but its share of the hotel market is small despite numbers rising from 6 in 1988 to 23 two decades later. In 2008, only 3.6% of all hotel bed spaces were in this category (*Dades Informatives*, 2008: Table 7.7).

The third and most important point is that following autonomia in 1983 the second boom of the tourism industry had become so important to the island's economy that it inevitably became absorbed into the local political processes especially those associated with the budget and fiscal policies. Party politics in the Balearic Islands, and in Mallorca particularly, is forged as coalitions. While there are right and left parties, none can afford strong ideological stances; in the search for power they constantly have to make compromises. The hotel owners, and more especially the large hotel corporations, became important players in the political system. The major hotel chains were able to operate outside the mainstream because of their size and power. They were able to direct negotiations with tour operators, the Govern Balear and even with Madrid. This tended to be a conservative relationship that sought to defend the free market economy, to (over-) exploit resources and energy and to influence resource and land-use planning to its advantage. Naturally, their political influence was directed at supporting those parties that had similar objectives. Perhaps, the picture has never been quite as bald as this as can be seen in the work of the first councillor for tourism, Cladera (1983–1993), who came from a hotel background but introduced some of the earliest and most influential planning legislation in 1984 and 1987 (Amer i Fernandez, 2006a,b: 51–54).

A number of examples are cited by Amer of this varying relationship between the hotel sector, political parties and policies for tourism in Mallorca. Under Gabriel Cañellas (the first President of the Govern Balear following autonomy, 1983–1995), intervention was kept to a minimum, following the tradition established under the Franco regime and its successors during the transition to democracy but increasingly in this period planning first became important politically. In *Autonomy and Society*, a publication of the right-of-centre party Aliança Popular, he argued that it was tourism that had created the wealth of Mallorca, a position reached through its own efforts and not those of the state but that the hoteliers should now recognise their responsibilities for modernisation and environmental management (Serra i Busquets & Company i Mates, 2000: 80–82). There was notable tension between hoteliers, represented by CAEB (Confederation of Business Associations of the Balearic Islands), and 'planning'. The first Councillor for Planning, Jeroní Saiz, acted as a moderating influence between the expansive and, some would argue arrogant, building industry and the Govern Balear. Cristofol Cladera, as the first Councillor for Tourism, had a background in the hotel industry and played an even more forceful part. The powerful Foment de Turisme advocated the expansion of tourist resorts into formerly undeveloped areas such as Cala Mondragó and the huge unspoiled beach of Es Trenc. The municipality of Santanyí certainly supported the former as did the political parties PP and UM at the Govern level. However, popular opposition was mobilised via a series of demonstrations, resulting in the Govern purchasing the land around the cala, eventually designating it a natural park in 1992 (Amer i Fernandez, 2006a,b: 67). Similar popular movements outside the ambit of party politics have been successful in resisting development at Es Trenc and the island of Sa Dragonera.

Other examples can be drawn from the history of golf course development in the 1990s. The establishment of ANEIs and the ARIPs was designed to give a measure of protection to rural landscapes and was initially supported by the hotel federation. As we have shown, golf courses have often included urban development in their specifications in order to make them more attractive to investors. The golf course proposals for Son Vida and Val d'Or were promoted as bringing to inland areas the kind of economic benefits originally confined to coastal tourism even though they would have contravened the landscape legislation. On being elected in 1991, the PP–UM coalition modified this legislation and despite opposition from popular movements such as GOB the developments were approved.

Two additional political elements can be seen at work, then: the municipality and pressure groups working via non-government organisations and popular demonstrations. The role of the former has been variable over time and although subject to varying ideological stances (free market vs. planning) has usually supported the 'the market' whereby municipalities obtain their 'fair share' of tourist developments in order to maximise local benefits from taxation, the sale of building permits and, of course, employment opportunities. Certain coastal municipalities such as Calvià, Andratx and Palma/Llucmajor have been the El Dorado of Mallorca's tourism wealth and others to the east and north coasts have similarly benefited, but the inland and mountain municipalities have seen the opposite effects for their local authorities. Alongside this spatial variability must be placed the conflict between the municipalities and both the Consell of Mallorca and the government of the Balearic Islands as a whole. Examples were given earlier of the opposition to support for beach protection measures. For a previous era, Rullan Salamanca points to the role that the central (Madrid) government had in trying to protect some coastal municipalities from over-development when many local authorities were simply too weak to resist the blandishments of resort developers and hoteliers; the 1960s witnessed an anarchic and random approach to planning in which the hoteliers appeared as more powerful than any elected representatives (Pack, 2006a: 173; Rullan Salamanca, 1989: 99–105). Picornell and Picornell (2002a: 102) remind us, of course, that municipalities are proud of their independence and want what is best for their citizens, a stance that has sometimes led to mitigating pleas that often followed alleged corruption. One of the ironies of this concern of municipalities for their economic and social well-being has been that the alternation of elected political parties at this level can result in one coalition supporting strict control when in power only to be succeeded by a corrupt coalition intent on reversing the policies clandestinely; Calvià, a municipality that contains about a quarter of Mallorca's tourism, has suffered especially in this respect (Rullan Salamanca, 2005). Many inland municipalities have been somewhat less than stringent in their application of planning legislation and have allowed building permits in order to secure second home developments in the hope of economic benefit (Salvà Tomàs & Binimelis Sebastian, 1993: 75).

The global economic crisis that began in late 2007 and continues today has had a profound effect upon the Spanish and Mallorcan economies. While irregularities and weakness in the banking sector became apparent in many countries – especially in Greece and the Republic of Ireland – in Spain the stress has been exacerbated by massive over-investment in the construction and house building sectors. In Mallorca, the problem has been somewhat less acute because the tourism economy proved more robust than expected so that investment in housing was at a lower level. Because demand was running at a high level for reasons of demography and household formation, fewer properties have remained empty.

However, the island's economy has had to bear its share of reductions in national public expenditure. This has resulted in the reorganisation of many government departments including those with responsibilities for tourism policy and planning, resulting in emphasis shifting away from the Govern Balear towards the island *Consells*. As Mallorca's Consell has a different political complexion from that of the Govern, it will be interesting to see if differing ideologies and political constraints will impact upon policy. In particular, how different will the relationship between the Consell de Mallorca and the interests of the powerful hotel business lobby be? At the time of writing, it is too early to judge (see comments in Chapter 7 and p. 188).

The Ecotax: A Political Perspective

Taxation can be seen as an example of the interplay between politics, the economy and business. In the previous chapter, the ecotax was cited as a possible means of counteracting environmental degradation. One of the most noteworthy of the conflicts in recent years between different political parties and their ideologies was the case of this innovative tax (Palmer & Riera, 2003: 665–674). Taxation of tourist activity seems an obvious target for fiscal policies in Spain generally and in the Balearic Islands in particular. As a major form of economic activity, it should yield a rich tax harvest. However, because of its central role and its sensitivity to overseas demand, tourism has usually been handled with kid gloves by the Spanish Treasury, exemplified by the fact that until recently hotel rooms received a much lower VAT levy at 44% of the general rate of 16%. (The rates were increased in the summer of 2010 to 8% and 18%, respectively, in response to the economic crisis.) The reasons for taxing tourism are threefold: the receipts could be considerable, such receipts could be used to fund tourism-related costs that are often higher than average costs and the returns greater and thirdly receipts could be used to offset environmental disbenefits (Gago *et al.*, 2009: 388). Palmer and Riera also point to the general effects of pricing tourist services too low (Palmer & Riera, 2003: 666).

The ideology that led to this tax was the product of two forces: the perceived need to raise governmental income to tackle the adverse environmental externalities that were created by tourists but the costs of which were borne by residents, and the political beliefs of a progressive coalition that was more redistributive in outlook. By the late 1990s, many in Mallorca believed that tourism was becoming unsustainable and its growth needed to be 'corrected' by fiscal measures in addition to the planning controls that had so often proved too weak (Picornell & Picornell, 2002b: 102). This may have been the view of the Pacte de Progrés and its supporters, but it was not one shared by the right-of-centre opposition parties in the Govern Balear, the hotel industry and not by the Madrid government of the Parti Popular led by Aznar. Naturally, the proposal stimulated a fierce debate between these opposing forces. Overseas tour operators – one of the most powerful lobbies in Mallorcan affairs – raised the ultimate threat of taking their business elsewhere, at a time when other resort regions were expanding and challenging the island's supremacy anyway. Local hotel groups believed the tax to be discriminatory and therefore possibly unconstitutional. They thought the revenue from such a tax was likely to be insufficient for its purpose. In addition, taxation in Spain, as in most countries, is notoriously difficult to direct to specific purposes, in this case environmental improvements. The press in the Balearic Islands for the most part supported the hotel industry's opposition as did many British newspapers and tour operators' organisations ABTA (United Kingdom) and DRV (Germany), all arguing that it could add substantially to the cost of holidays (*The Daily Telegraph*, 23 February 2002).

Eventually, the Spanish courts upheld the Govern's decision, but the tax was not to last long partly because of the range of the forces against it and partly because of a change of government in the islands in 2003. The ecotax did not aim to recover all the monetary costs of the environmental and resource disbenefits (Palmer & Riera, 2003: 669), but it would have eventually raised 60 million euros each year by levying a charge of 0.25–2 euros on each tourist per night according to the type of accommodation. This figure was substantial, being equivalent to 5.6% of the Balearic budget in 2002 or 45.7% of the Mallorcan Consell's budget or 24.5% of the annual precept sent by Madrid to the autonomous government of the Balearic Islands (Mallorcaweb, www.mallorcaweb.com, 19 January 2002; accessed 10 March 2010). When it was removed in July 2003, the argument that it was discriminatory was used; it applied only to tourists staying in official accommodation and not to those staying with friends or residential tourists. In colder political terms, it was the opposition of the Parti Popular and Union Mallorquin and their friends in the Federacion Empresarial Hotelera de Mallorca that killed it off.

Internationalisation of Mallorca's Hotel Business

The history of the development of the hotel business in Mallorca illustrates that from the very beginning the sources of capital, labour, investment and innovations were not confined to the island's shores. Even the building materials and the architectural styles were largely imported. What was Mallorcan was the organisation of vacations for the mass of tourists: the particular combination of overseas tour operator, airline and island transport with local hotel management to produce large numbers of holidays for large numbers of people at relatively low cost but with an acceptable level of service, all supported by a compliant government in terms of tax regimes and planning controls. This is the main characteristic of the Balearic Model referred to in earlier chapters. It provided widespread opportunities for small hoteliers to join this process, later to experience rationalisation and the emergence of a corporate sector of well-known brands. As long as the number of tourists continued to rise year-on-year (and for most of the last 70 years that has been the case despite brief periods such as the early 1970s, 1990–1991 and 2007–2010), considerable profits and surpluses became available for reinvestment or translation into other forms of economic activity. However, for the most part, the Mallorcan hotel industry has sent its surpluses into overseas tourism, particularly to the Caribbean and Latin America.

The process of structural rationalisation and horizontal integration, accompanied by expansion overseas, has resulted in the domination of the hotel industry in Mallorca by four large corporations who between them control more than 640 hotels worldwide with more than 180,000 rooms (Table 8.3) (Serra, 2009: 127).

All this originated from a series of family-run hotels, characteristic of the Balearic Model in its early stages. Two companies may be used to illustrate the processes of change.

Sól Melia, with about 330 hotels globally, began in 1956 with Gabriel Escarrer's first establishment in Palma, which eventually grew into the small Balearic chain, Hoteles Mallorquines Asociales (Table 8.4). By the 1980s, this had expanded into the Peninsula and the Canaries, built its first hotel in Bali (Philippines) and acquired the Hotasa chain. In 1987, it merged with the Melia organisation, with the latter concentrating on business hotels and the Sól part remaining with conventional tourism (Pellejero Martínez & Martin Rojo, 1998). In 2000, it acquired the Tryp

Table 8.3 Balearic hotel chains with presence in Latin America, 2006–2007

Chain	Rank in Spain in 2007	Hotels (n)		Rooms (n)		Countries (n)	
		2007	2006	2007	2006	2007	2006
Sól Melia	1	218	167	43,625	44,346	32	26
Barceló	3	85	73	23,183	19,903	13	13
Riu	4	61	63	22,848	22,167	17	17
Iberostar	5	63	62	21,258	17,228	14	12
Bahia Principe	8	11	6	6,453	4,757	3	2
Fiesta	9	11	11	4,358	4,358	3	3
Sirennis	11	5	4	2,811	1,770	3	2
Hotetur	15	10	11	2,465	2,710	3	3
Total		464	397	127,001	117,239		

Source: López Nadal and Maturana Bis (2008)

Table 8.4 Sól Melia group

Chain	Country	Hotels		Rooms	
		2007	2006	2007	2006
Sól Melia	Costa Rica	3	2	839	731
	Cuba	26	23	10,126	8,476
	Mexico	10	10	3,376	3,395
	Panamá	1	1	305	287
	Republic Dominicana	5	4	2,403	2,175
	Venezuela	2	1	664	664
	Argentina	3	2	328	182
	Brazil	20	21	4,217	5,154
	Perú	1	1	180	180
	Uruguay	1	1	74	74
Total		72	65	22,512	21,318

Source: López Nadal and Maturana Bis (2008)

chain and so today it operates in 30 countries and is the largest hotel group in South America and the Caribbean as well as Spain.

The Barceló group began life as a coach company in Felanitx in 1931, turning to the travel agency business in 1954 and opening its first hotel at the peak of the first boom in 1961 (Table 8.5). Like Sól Melia it expanded initially into the Peninsula, with its first overseas hotel in the Dominican Republic in the mid-1980s. Since then it has expanded into many parts of Latin America, Turkey, the Philippines and the United States via a series of takeovers including of Turavia in Spain and Crestline Capital in the United States. Although it has developed some urban operations, it remains essentially a sun, sand and sea outfit in the mass market (Serra, 2009: 128).

Both these multinational corporations began by working closely with tour operators who were responsible for distributing hotel capacity between competitors. They were the Mallorcan hotel end of a European system of low-cost holiday provision. Once the decisions were made to expand overseas, these companies by-and-large followed their competitors into emerging geographical markets, behaving similarly to other realms of economic activity. Costs were usually lower overseas, and local governments were keen to encourage inward investment rather like Spain had been in the late 1950s. López Nadal and Maturana Bis (2008) have described this investment into Latin America as a *reconquista empresarial*, Pizarro-like with colonial overtones. Between 1993 and 2008, one third of all hotel investment in Latin America came from the Balearic Islands, with

Table 8.5 Barceló group

		Hotels		Rooms	
Chain	Country	2007	2006	2007	2006
Barceló	Costa Rica	5	5	871	871
	Cuba	3	3	1,464	1,464
	Mexico	10	10	4,044	3,681
	Dominican Republic	10	9	3,850	3,055
	Ecuador	1	1	95	95
	Nicaragua	1	1	283	283
	Uruguay	1	1	75	75
Total		31	30	10,682	9,524

Source: López Nadal and Maturana Bis (2008)

greater concentrations in some countries, notably the Dominican Republic and Mexico. This was the result of both push and pull factors. Amongst the former elements were the fact that Mallorca was approaching the end of the Butler cycle by the 1990s, the ousting of the 'friendly' PP and the introduction of the ecotax, the ready transfer of Mallorcan know-how to new environments and the further vertical integration of the large hotel chains into areas such as travel agencies and construction. The pull factors include the obvious historical and linguistic ties of Mallorca with certain Latin American countries, the liberalisation of the World economy generally that encouraged inward investment at this time and, of course, the physical resource base of parts of these countries and islands leading to the familiar sun and sand developments onto which could be grafted modish ecotourism. Cuba was something of a special case where a revived tourism industry encouraged inward investment following the decline of Soviet and Russian influence (López Nadal & Maturana Bis, 2008). Clearly, much of this new international business followed the original, highly profitable Balearic Model for mass tourism, with the hotel companies remaining in family control. As Mallorca and even Spain became 'full' with rising negative externalities and growing popular opposition to further expansion in the islands, development overseas increasingly became an attractive investment opportunity. It was also a means of challenging competitors' expansion in new regions such as the Middle East, Eastern Europe and Asia Minor. Profit was a major driving force but not the only one. Recently, these global businesses with Mallorcan origins have been strongly criticised by many islanders for investing overseas rather than facing up to their (largely environmental) responsibilities at home (Serra, 2009: 130–132).

We have also seen earlier that the larger hotel chains are those that adopt new technologies more readily. Overseas, any change that may have affected the labour market was easier to introduce. Another tactic has been to enter into alliances with tour operators, with Barceló, for example, working very closely with the British company First Choice and Rui with the German company TUI. Overseas, some chains, notably the Sól Melia group, have moved away from family ownership towards a stock exchange quoted company that contracts out or franchises many of its operations, demonstrating the convergence of the hotel industry with other forms of economic activity in the way it organises its business.

Some Conclusions

This chapter has attempted to illustrate the ways in which the hotel business of Mallorca has shifted over the last 100 years. From beginnings with luxury hotels for the travelling bourgeoisie of northern Europe and north America, it has moved through eras of dependence upon the

Spanish market, the teeming masses of urban, industrial west Europe seeking low-cost accommodation in comfortable hotels next to the beaches and on to the rationalised, multinational operations of a small number of massive chains. Although over much of this century it retained a particularly Mallorcan flavour to this process, it is now evident that the island's hotel system has become part of a much wider global pattern that has rather different social and economic objectives from those of the founders in the early years and the family businesses of the first boom. Economy, society and polity in the Mallorcan tourism industry have become inextricably linked.

Chapter 9
New Markets and Diversification

Introduction: Dependence

Almost as soon as the Balearic Model reached the take-off stage, Mallorcan-based hoteliers and entrepreneurs became aware of the competition from other, usually Mediterranean, geographical regions that could, and probably would, begin to develop similar resorts based on similar economic and environmental principles. Malta, Greece, Cyprus, Morocco, Tunisia and above all the coastlands of the Spanish peninsula all learned from Mallorcan experience. Mallorca was already heavily dependent upon two markets: the British market and later the German one. Because of the diversity in the structure of the hotel industry, Mallorcan response initially to this competition and dependence was not coordinated although the national government, especially under Fraga's direction in the 1960s, was able to make the industry more aware of what was happening elsewhere (Pack, 2006a,b: 102). The massive increase in the scale of activity in the first 'boom' made the threat of competition seem remote, but the period after the price shock of the 1973 energy crisis brought a new focus to two areas of activity in order to maintain and preferably improve competitiveness: new markets for the Mallorcan product and its potential for diversification. The inexorable rise in tourist numbers since the early 1960s, despite the anxieties of the mid-1970s, early 1990s and now the end of the first decade of the 21st century, has been remarkable (Figure 4.1). Some of the innovations in the last 20 years are set out below. The changing relations between the hotel industry and island polity were discussed in Chapter 8.

New Markets

In 1996, this author suggested that one new source of potential tourists to Mallorca was likely to be the countries of the former Eastern Europe and the Soviet Union (Buswell, 1996: 336). As the strict control on overseas travel was relaxed with the fall of communism and the rise of more democratic forms of government and as many of these countries joined the European Union and even adopted the euro, it might have been expected that the Mediterranean, and Mallorca in particular, might prove to be an attractive tourist destination. The more northerly nations such as the Baltic States and Poland and the landlocked countries such as the Czech and Slovak republics were identified as more likely to supply more tourists as was the new Russia. As disposable incomes slowly rose in these countries and leisure time increased, it was hoped that more of their citizens would find their way to Mallorca's beaches. Numbers have risen over the last 20 years, but their ratio to tourists from other countries has increased only marginally. For example, Russian tourists to the Balearic Islands increased from 10,000 in 1994 to 36,000 in 2008 (+ 360%) but as a proportion of all visitors, this increase was only from 0.13% to 0.29% (+ 220%) (*Dades Informatives*, 2008: Table 1.3). Certainly, Mallorca now receives tourists from a much more diverse range of countries than in the 1960s, but the numbers and their proportion remain dispiritingly small. Finns, Portuguese and even Luxembourgers outnumber Russians with their enormous population resources.

One remarkable change – at least remarkable to bourgeois Spanish mainlanders – has been the increase – absolute and relative – in the number of tourists from the Peninsula who now account for about 18% of visitors, a proportion not seen since the late 1950s and rising from only 14% in 1990 (Llibre Blanc, 2009: 65). By 2008, about 1.7 million Spaniards came to Mallorca out of a total of 9.9 million, and like foreigners, mostly by air (*Dades Informatives*, 2008: Table 1.3). This is partly explicable by marketing efforts, but much more significant has been the socio-economic changes in the Spanish population regarding affluence and leisure time together with the changing perception of Mallorca by *peninsulaistas*. The mass tourism experience is now one that many Spaniards want to participate in, with Mallorca no longer seen as remote and provincial, the destination of only British and German lager louts. Similarly, of course, affluent Mallorcans now holiday aboard especially in the low season. The generous allowances to Spanish pensioners have also been responsible for an increase in mainlanders coming to Mallorca in the winter months.

Additional numbers from 'new' sources have certainly been coming, but with the exception of mainland Spanish visitors their ratio to total numbers has increased only marginally if at all. Some formerly traditional counties such as France have fallen away; Mallorca has relied on continuing increases from Britain and Germany. Table 5.3 shows their relative distribution in the major resort regions.

Diversifying the Tourism Product

Historically and economically, the market for tourists in Mallorca has been the mass market and the two-week package holiday in a three-star hotel located in a 20th-century coastal urbanisation, highly seasonal and climate and weather dependent. Between the late 1950s and the early years of this century, this model has provided consistent growth in tourist numbers year-on-year. To many observers, it seemed that this expansion could only continue as European disposable incomes rose and leisure time increased. The prices of package and independent holidays fell in real terms thanks to falling transportation costs by air and the willingness of Mallorcan hotel groups to discount prices whenever numbers appeared volatile. To German and British holidaymakers, the attractions of 'Majorca' held fast; after 1989, the potential of tourists from the former East European bloc did not really materialise.

However, this is something of a simplistic picture. The growth curve was never consistent – witness, for example, the downturns of the mid-1970s (quintupling of world oil prices), the early 1980s (British deindustrialisation), the late 1990s (German economic stagnation) and the recent global recession 2007–2010 (Figure 4.1). The crucial variable became market share. While the Balearic Islands were fairly consistent in their share of overseas tourists coming to Spain as total numbers grew, it was becoming clear by the mid-1990s that their share of world tourist numbers was declining as Europeans looked to other parts of the Mediterranean – Greece, Cyprus, Malta, Tunisia, Morocco and Turkey – and to North America, especially Florida for the British. Secondly, by about the same date, the pressures of tourism on the environment of Mallorca were becoming less tolerable – to both visitors and Mallorcans alike. The nature of this pressure has been discussed in Chapter 6. For any response to these two factors, it was clearly best if they were seen as being interrelated if not actually interdependent. Mallorca needed to not only retain its market share (relative numbers of visitors) but also manage, or better still, to reduce their impact on the island's society and its environment. In socio-economic terms, we have seen that a major problem from the 1960s has been the seasonal nature of the industry and its impact on the labour market. In the early part of this century, this was compounded by its changing nature whereby as more and more people worked outside the tourism industry, labour shortages began to be felt, especially in lower-grade occupations in nearly all areas from the health service, local government and construction, to tourism itself. This led to a relaxation of immigration policies under Zapatero and a very considerable influx of workers into Mallorca. In 1998, only 4.7% of the population of the Balearic Islands had been born overseas, but a decade later, this figure had risen to 20.8%, the highest proportion of any Autonomous Community of Spain (Diario de Mallorca, 21 June 2008; INE, 2008, see statistical sources). Immigration is now on such a scale that Mallorca's economy probably cannot support such numbers without a return to the pre-2007 levels of tourist numbers.

The urge to diversify the tourist economy as a policy objective has been motivated by various factors, including the need to maintain market share by improving the island's attractiveness in the broadest sense, to reduce its environmental impact, to try to reduce seasonality by distributing numbers more evenly throughout the year and to attract its share of the so-called new tourists. Nearly all these are variations on the

Balearic Model as seen from the consumers' point of view. The Mallorcan authorities hope to achieve this diversification while at the same time maintaining income from tourism in real terms. There is also pressure to reduce the dependence of the island on the economic and social monoculturalism that currently attends mass tourism. Some of the implications are discussed in Chapter 10, but the challenge of diversification has been confronted since the late 1970s. To date, however, it has been met largely on an ad-hoc basis. Statements in the second Llibre Blanc (2009) imply that only a coalition between political, business and environmental interests will be equipped to succeed.

In this chapter, two broad factors are implied in these policy shifts. The first is that if the structure of the industry is changed to accommodate a more diverse range of tourists and at different times of the year, then the income derived must at least equal current income levels, preferably adding to them. If there is a net increase in numbers as a result (put simply, 'new' tourists plus mass tourists), then the externality costs (especially environmental ones) must not increase unless Mallorcan gross national product increases sufficiently to pay for them (environmentalists would obviously argue against this). The second is concerned with the diversification of Mallorca's economy as a whole. Other parts of economy – agriculture, manufacturing and the tertiary and quaternary sectors – will need to be able to attract capital and labour, invent – or more likely innovate – new forms of activity and products and at the same time be accommodated within the island's social, economic and physical environment without adverse effects.

From about the mid-1980s, the ad hocism referred to has had some positive effects in both areas. Tourism is less seasonally concentrated, new forms of tourism have become evident, especially those associated with leisure and recreation, the socio-economic profile of tourists has been broadened and some additional geographical zones have been developed, notably away from the crowded coasts. In the non-touristic economy, there has been less success because of its ultimate dependence. Agriculture has been declining but diversifying if only marginally. Manufacturing has behaved in a similar fashion. It is in the service industries where most structural change has taken place, but much of this has been in the public sector, at all levels of government and in public services such as education, health and transport. Perhaps only in banking and financial services and construction has employment and output increased in the private sector, and it is often difficult to disentangle this from the performance of the tourism industry.

It is against this background that attempts to diversify the tourist economy have to be examined. A number of sectors ranging from the conference trade to hiking and walking have been selected for analysis.

Conferences

Like many other tourist regions, the Balearic Islands, and especially Mallorca, identified the conference trade as a potential source of diversification. Amongst the reasons that are said to recommend its candidacy are the well-developed infrastructure for all tourists that can apply equally to the conference delegate, the marked seasonality of conventional tourism that creates an opportunity for additional customers in spring and autumn and the economic benefits that might derive from this trade as convention participants are perceived as high spenders and can bring prestige to a city or centre.

Conferences (Sp. *reuniones*) consist of three types of meetings: congresses, which tend to be relatively open and appeal to a wide audience usually with large numbers spread over a few days; conventions, which are more client specific (e.g. travel agents), which again tend to be large but shorter in length, and finally corporate functions, often seen as rewards or incentives for employees or specific client groups but usually short, often not involving overnight stays. Many participants, of course, see elements of all of these in the attraction of Mallorca as a conference destination.

This kind of tourism is now big business especially in the United States but nonetheless Europe commands nearly 60% of the world trade. In 1996, there were 3500 international conferences, a figure that had risen to more than 8200 by 2009. Worldwide Spain now ranks third behind the United States and Germany. However, in terms of number of participants, France has a quarter of a million to Spain's 180,000. Vienna, Barcelona and Paris are the main convention centres, with Madrid having only one third their number of conferences. Medicine, science and technology dominate the sectoral breakdown with nearly half of all international conferences worldwide. As in so many other areas of business activity, there has been a pattern of concentration into larger gatherings at the expense of smaller ones, and in fewer centres. There is a plethora of small, more specialised meetings nonetheless, but their impact tends to be limited and highly localised and not geographically mobile over long distances (ICCA Statistics Report, 2008).

There are cultural differences in the nature and location of conferences and so medicine, for example, has large conferences held primarily in high-quality hotels, especially in the United States, a reflection perhaps of doctors' high incomes and sponsorship by powerful pharmaceutical companies. Other important groups use specially built convention centres and auditoria such as political parties and trade unions, whereas lowly academics are often confined to the meagre facilities of universities' lecture rooms and student residences. All of these types have a mixture of large-scale plenary sessions and smaller lectures, seminars and discussion groups, which make demand on hotel or conference spaces in different ways but both require a combination of spaces.

Within Spain, the Balearic Islands rank relatively low as a conference destination, with less than 2% of the country's meetings and an even smaller proportion of participants; Madrid's figures are approximately 20% under both headings. Catalunya and Andalucía rank second and third as destinations. Very few of the conferences and meetings held in Mallorca are truly international; most of the organisations come from the Peninsula. Only Mallorca amongst the islands has the larger four- and five-star hotels required by international conferences and the larger specialist auditoria. International accessibility by air is good only within Europe with few, if any, direct flights to North and South America, Africa and Asia/Australasia. And even here, the frequent flights to Germany and the United Kingdom are dominated by no-frills airlines not always appealing to the well-heeled international doctors, scientists or business-men/women (INESTUR-CAEB, 2006). There is also a conflict of use between the spring and autumn conference trade that might come to Mallorca and the already diversified middle-class or elderly conventional tourists the island has been quite successful in attracting to its better hotels. Some of the auditoria in Mallorca, such as the Palau de Congressos in the Spanish Village in Palma and the Auditorium de Palma, are now aging having been built in the 1960s. Conference centres outside Palma are small in numbers and size and unlikely to be attractive not having the extramural attractions of the island's capital. To offset these limitations, the Govern Balear has commissioned a new Congress Palace in Palma to house more than 2000 delegates and to include a hotel with a capacity of 600.

The result of this honest but rather depressing picture is that while Mallorca and Palma may appear to be an attractive venue for conferences, the truth is that Palma remains a provincial city, even within Spain, unlikely to figure more prominently on the map of the world's conference tourist business.

This trade in Mallorca currently attracts about 180,000–190,000 delegates each year to just over 3000 'events' of which fewer than 2000 are conferences or trade fairs. The majority of these are held in hotels (92%) in spring and autumn, mostly coming locally from within Spain and from Germany and the United Kingdom. Although the economic impact has been increasing to more than 200 million euros annually, without the North American trade it is difficult to see this pattern changing. It is a business that is sensitive to the world trade cycle (Mallorca Daily Bulletin, 14 April 2010).

Cruise Tourism

A second area where Mallorca has hopes of expanding its diversified tourism product is the cruise ship industry. In an earlier chapter, it was noted that in the 1920s and 1930s cruise tourism was an important ingredient of tourism generally in Mallorca. Mulet (1945) recorded 360 passenger ships calling at Palma in 1935–1936, with over 5000 disembarking. In pre-airline days, passenger ships between Europe and Africa and Asia crossed the Mediterranean, linking colony to homeland. Freighters on similar routes often carried fee-paying passengers. Once aircraft replaced ships, this pseudo-cruise market collapsed. Slowly in the post-war period, particularly in the late 1980s and 1990s, the number and size of cruise liners grew and their routes became more dispersed as cruise tourists looked for more and more exotic locations. In this, cruising was not so different from general tourism worldwide. For the American market, the Caribbean remained dominant from the 1930s onwards. For Europe, the Mediterranean predominated, harking back perhaps to an era of more leisurely travel featuring Swans Hellenic Tours and an interest in the 'Classical'. Clearly, the specialised cruise liners of today offer a rather different tourism product from the passenger and freight ships of the 1930s, even 1950s. Mallorca, in the generosity and expansiveness of Palma harbour, obviously has at least one resource attractive to the cruise trade.

By 1970, there were about half a million cruise passengers on the World's oceans, a figure that was to rise to 13 million by 2009, of which more than 10 million originated in North America. From 1996 to 2002, European cruise tourists increased by 70% and US cruise tourists increased by 48%. By 2005, it was estimated that there were 250 vessels plying their trade with a capacity of about 13 million passengers: 37% of this trade was directed to the Caribbean, 15% to the Mediterranean and 5% to the Baltic. The United Kingdom supplies nearly 1.5 million cruise tourists, nearly 600 of whom visit the Mediterranean, Germany about 860,000 and Spain about 450,000 (IRN Research, 2009).

The nature of cruising has not changed substantially from the 1930s. Much of it appeals to a luxury market with high levels of service (e.g. ratio of crew to passengers), quality and quantity of food and drink consumed each day and a wealth of diversionary on-board activities ranging from hairdressing and beauty treatment to health clubs, swimming and dancing. While originally aimed at a closed world of pampered luxury for high-income, middle-class professionals, in recent years it has moved towards a more democratic market. While not a 'mass' market, the industry has embraced tourists not traditionally seen as cruise material, a social process made possible by the increase in size of cruise ships that can accommodate various classes of passengers who, while not actually physically segregated as they were in a previous era, can and do crystallise into different social groups (Cartwright & Baird, 1999: 128; Mallorca Daily Bulletin, 6 June 2008).

Against this background what patterns and processes are emerging in Mallorca's cruise tourist business? Firstly, one of its main attractions from the industry's point of view is derived from the island's location in the centre of the west Mediterranean basin, accessible to many other ports-of-call in the area such as Barcelona, Girona, Naples, the North African ports etc. It is also an island set in a relatively calm sea with few of the Caribbean's hurricanes or the Bay of Biscay's storms; passenger comfort is a major consideration. Secondly, Palma is a busy commercial port not simply a cruise destination and so it is well equipped to deal with all aspects of ship service. Thirdly, Palma is very well connected with many of west Europe's cities by air, making it a suitable place for fly/cruise operations and hence as a base for operations where cruises can start and finish. The attractions of Palma and Mallorca for disembarking tourists will be dealt with below.

Currently, relatively few world cruise tourists come through the Balearic Island ports; for example, in 2008, Miami had more than 4 million cruise tourists coming through its port, whereas Palma had 1.3 million, a substantial figure nonetheless with only Barcelona exceeding it in the Mediterranean. After a rise in the 1960s and 1970s to about 220 ships per annum, numbers declined to an average of 130 in the 1980s. The 1990s saw the beginning of a marked increase in activity to a peak in excess of 700 in 2008. Passenger numbers rose from half a million in 1998 to the 1.3 million quoted above. In October 2008, as many as nine cruise liners docked in Palma in one day decanting more than 8000 *cruceristas* (Diario de Mallorca, 11 October 2008). As many as 19 or 20 different companies now use the islands' ports (overwhelmingly Palma, Mallorca) employing some of their largest vessels (more than 70,000 tons) belonging to P&O, Holland America, Celebrity Cruises and Cunard (Table 9.1).

Table 9.1 Mallorca: Cruise ships and passengers (1964–2008)

	Year			
	1964	*1997*	*2006*	*2008*
No. of ships	246	606	649	717
Passengers	63,175	574,739	1,060,060	1,314,101
Passengers/ship	257	958	1,633	1,833

Source: Dades Informatives (2009: Table 9.2) and own calculations.

There tends to be seasonality with cruising in favour of a long central period from May to October when some 35% pass through and only 2% between December and March. This does tend to defeat one of the supposed benefits of cruise tourism for Mallorca, namely lengthening the season. However, this appears to be changing with a decline in recent years in the high season and a rise in the mid seasons.

Cruise tourists are mainly middle class, professionals and managers, British (36%–37%), with slightly more males than females and with half of all passengers over 50 years old. Besides the obvious attractions of relaxation, good food and drink and getting to know other passengers, most record visiting different countries and cities and experiencing their cultures as principal motivations, and so the overall itinerary of a cruise route is important. So, what is it about Palma and Mallorca that is attractive as a port-of-call? Firstly, the majority of travellers opt for half a day of independent excursions rather than as part of an organised group. Unfortunately, few data exist as to what those disembarking actually do when ashore. Local observation suggests few venture beyond Palma, its shops and tourist 'sights'. This is in contrast to the experience in the 1930s when passengers joined car and charabanc trips to further parts of the island to visit its caves, its quaint railways, monasteries, mountains etc. (Mulet, 1945). Spending surveys, however, give some additional clues with eating out significantly more important (in fact, costly) than visits to museums, galleries and using local transport. Cruises average about 10 days with a per diem spend per tourist of 140 euros. Palma is increasingly a base destination, that is, a point of arrival and departure for a cruise. In 1998, 28.2% of cruises using Palma were of this type, but by 2008, this figure had increased to 36.4%. The majority of these passengers will arrive and depart by air. Passenger numbers over the same period on such cruises rose from 39.8% to 50.4% (*Dades Informatives*, 2008: 109). These data are important because normally cruises rarely dock for more than 1 day, but the average stay in Palma has been estimated at 1.5 days with an average 'spend' while ashore of 55 euros each day; Americans spend significantly more. From this, some rather crude estimates of the economic impact of cruise tourism on Mallorca have been made to show a figure of 65 million euros in 2004 rising to 90 million euros in 2008 (INESTUR, 2005: 144). These figures include direct and indirect spending by visitors. In this way, cruise visitors probably add significantly to urban tourism in Palma (see below). The ships also source more of their requirements locally when in a base port, but there are no estimates of the costs and benefits of this to Palma. The benefits of passenger and provision spending (fuel, food etc.) may be offset if docking fees are set low to attract liners.

As with other examples of diversification, there appears to be some benefit from cruise tourism but until some calculations of costs and benefits are made it is difficult to know precisely what they are. No mention has been made of the environmental impact of cruise tourism, but some research suggests that it can be considerable especially through CO_2 emissions (see Chapter 5).

Golf

Golf is perhaps a more logical development in tourist activity than the two diversifying products cited above. It is after all an established leisure activity but transferred to a different and perhaps more challenging environment. In 2002, there were 60 million golfers worldwide operating on about 32,000 courses (Palmer, 2004: 123). It has limited geographical and socio-economic characteristics being essentially a middle-class sport confined to North America, Western Europe and limited part of Asia.

The first course was constructed on Mallorca near Alcúdia in the 1930s, with the first modern course dating from 1964 at Son Vida. Although Spain has produced notable golfers in recent years, few, if any, have been Mallorcan; compared with football, athletics, tennis and sailing, golf is a more alien sport, a plantation from abroad. Within the Balearic Islands, the game is confined almost entirely to Mallorca with 20 courses; Menorca has only one and Ibiza two. The geographical distribution of courses is strongly linked to two factors: the most popular coastal tourist areas – there are none in the interior of the island – and close proximity to Palma and its airport; half the courses are in Palma, Calvià, Llucmajor and Andratx municipalities, suggesting that it is seen as a 'metropolitan' activity. Its contribution to spatial diversity is limited, tending to compound the geographical patterns of tourism generally (Figure 9.1).

Rather than the windswept links of eastern Scotland, golfers are increasingly attracted to Mediterranean surroundings. In this sense, it too shows seasonal tendencies, with March, April and October the most intensive months, which reflect seasonal deterioration in the home countries too. In the truly low season, golf holidays decline quite markedly and so their contribution to temporal diversity is limited (INESTUR, 2005: 115).

In landscape perceptual terms, one of the curious characteristics of Mallorcan golf courses is that they are designed for the most part to resemble those elsewhere in Europe and North America. Indeed, most of the island's courses have been designed by well-known American and European designers. As such, their appearance in the landscape is almost entirely foreign. They have as little reference to Mallorcan landscapes as do the multi-storey hotels to traditional Mallorcan architecture. This is true too of their physical appearance: sandy bunkers, neat closely cropped drives and putting surfaces, water hazards and often exotic

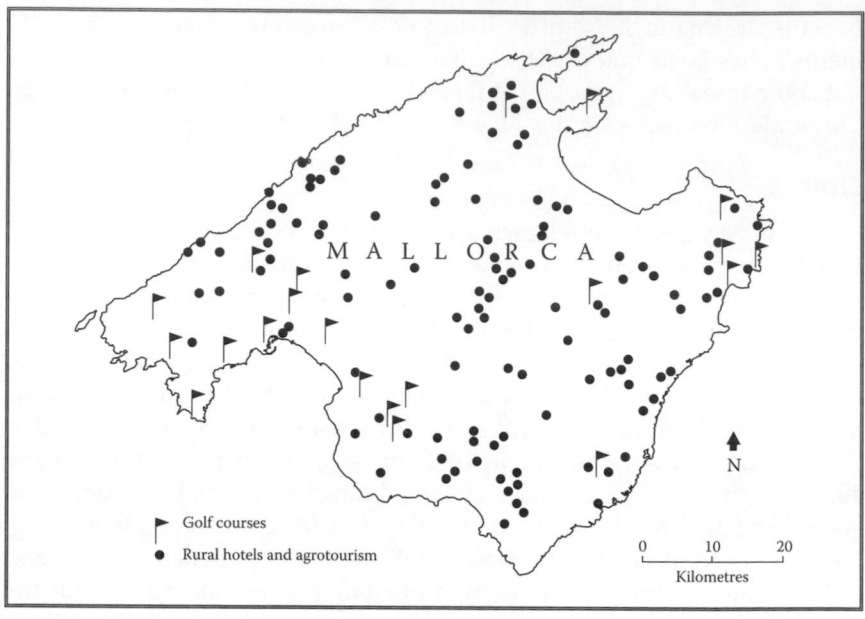

Figure 9.1 Map of golf and agrotourism areas

trees such as eucalyptus. These landscapes are the product not only of foreign designers but because of the use of that model, consume enormous quantities of water. It has been estimated that an 18-hole course can consume somewhere between 3000 and 10,000 m^3 of water per day, enough to supply a small-sized town (Grimalt Gelabert, 2002: 289; Standeven & De Knop, 1999: 248). However, since 1988, golf courses in Mallorca have been obliged to use recycled grey water. Unfortunately, in the high season, this has a tendency to smell, hardly conducive to an expensive round!

A second characteristic, by no means confined to Mallorca, is their development as part of wider real-estate concerns. The reciprocal symbiosis between golf and real estate is probably illusory, but few golf course developers could obtain sufficient returns on their capital investments without this. So far, Mallorca has avoided the 'golf city' complexes of Murcia, Almeria and Andalucía, but nearly every course has its attendant house-, apartment- and hotel-building with associated restaurants and 19th holes. This reinforces the class-based nature of much of the sport, something perhaps to which the authorities are not averse, as part of a policy of diversification.

In 2008, more than 112,000 golfers and their companions stayed in Mallorca, on average for about 10 days, each spending over 200 euros per diem and so golf's contribution to the economy is approximately 183 million euros annually (*Dades Informatives*, 2008:113). For many Mallorcans, the key question is whether or not this figure is enough to offset the environmental and social costs of this kind of activity. Increasingly, there appears to be a shift in the nature and organisation of golfing holidays, with more people making bookings via the Internet and no-frills airlines. Is this a move towards mass tourism and golfing (Garau-Vadell & de Borja-Sole, 2008: 20)?

Urban Tourism: The Case of Palma

We noted in an earlier chapter that the first tourists to Mallorca who came by sea nearly always confined their attention to Palma and the district surrounding the city. The first luxury hotel (the Gran Hotel, 1903) was located in Palma. The expectation at that time was that the visitor would be primarily interested in the cultural and historic resources of Mallorca: sun, sand and sea came later. Where earlier visitors had spent most of their holiday located in the city plus excursions, the contemporary tourist sees the city as a peripheral location during the 2-week stay, a diversion or distraction from the beach and the coastal resort, the place for an occasional visit.

It seems possible to postulate two 'Palmas': that seen and used by the mass tourist and that perceived by the off-season visitor (Figure 9.2). Formerly, the attraction would have been one of cultural contrast in, for example, architecture and townscape (O'Hare, 1997). The 20th-century rectilinear streets and concrete and glass hotels and apartment blocks of the modern Mallorcan resort have only limited appeal, if they enter the consciousness at all, whereas the historic core of Palma offers diversions in its many narrow streets and stairways, its renaissance townhouses and palaces and its many medieval churches, especially La Seu, the great Gothic cathedral that stands above the bay. However, we have already shown that Mallorca is still essentially a venue for mass tourism and the package holiday and it seems legitimate to query the attraction of the historic resources of the city of Palma for the 'typical' tourist. Is it not more likely that the city has other, more culturally relevant attractions to offer? In particular, shopping, eating and drinking and the eponymous 'night life'? Even though many day-visitors may wander the medieval streets and plaças, their 'gaze' is much more likely to be directed towards the shop window and the terrace café than the *moderne* facades of Calle Colon. The environment in which much of the perambulations or open-top bus tours takes place is perceived by visitors as much less heritage than consumerist. The 'placeness' of Palma is not so much perceived as culturally or historically contrasting to 'home' but as all-to-familiar sights located in an historical setting (Urry, 2002: 115). The important

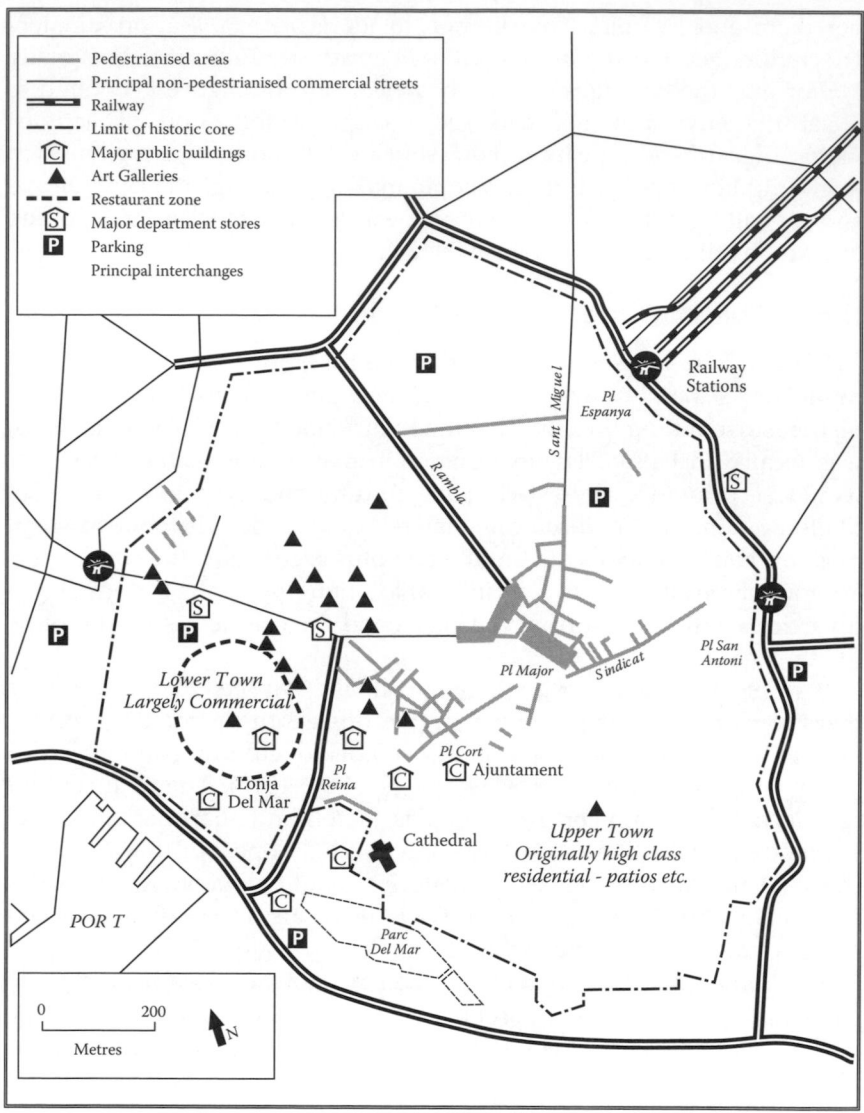

Figure 9.2 Map of Palma town

signifiers are not the townscape and the architecture but the department store, the myriad of shops, the motor traffic, the pollution, the over-crowding. Just as the appearance of the resort has become a familiar locale, the city of Palma is perceived as familiarly 'urban' – not so very different from Manchester or Düsseldorf or at least as filtered through those place-particular images.

Over the last three decades, the city fathers of Palma have endeavoured to alter the appearance of Palma conserving the best in architecture and townscape but often through the contentious process of gentrification. This is especially true in the core of the city where vast sums are being spent by developers with *ajuntament* and Consell support converting historic buildings and in some cases parts of whole quartiers such as Sant Pere and Calatrava into expensive homes, offices and boutique hotels. The relationship between conservation and tourism has not been well explored, but its relevance in the case of Palma is probably to the second group of tourists identified above and to which new policy initiatives are primarily being directed. As Ashworth and Bruce (2009) recently reminded us:

> ... insatiable, unsustainable tourist behaviour means no location can rely on innate attributes to keep tourists engaged with its attractions: all have to work to attract and if aiming for 'less unsustainable tourists' must seek ways of immobilising, de-mobilising or at least temporarily slowing down their guests. (Ashworth & Bruce, 2009: 301)

Urban tourism is seen by most as part of the diversification process in mass tourism resort areas. Like so many other examples of this process, the broad aim is to offer attractions to the mid- and low-season tourist, helping to flatten the seasonal curve while adding to the total sum of the of tourists' experience while visiting the island. For the local economy, its main effect is to extend job opportunities and hotel occupation rates in a place characterised by its extreme seasonality.

Research has shown, however, that success in urban tourism depends to a great extent on the uniqueness of the urban experience on offer (González Pérez, 2008) and while Palma has a fascinating history and topography (Buswell, forthcoming) it does not have the caché of a capital city such as Madrid nor the identity of a Prague, a Barcelona or a Dublin. It certainly possesses a high-class retail environment and an improving gastronomic infrastructure, but its artistic cultural attractions are limited. There is no world-class gallery, opera house or music venue to attract the well-heeled middle-class visitor despite considerable investment in the Solleric, Baluarde and Miró galleries and the renovation of the Opera House. The Auditorium on the Passeig Maritime is now very dated and in need of refurbishment as has already been noted in relation to the conference trade.

Amongst the improvement thought necessary to improve urban tourism in Palma, INESTUR has recently identified the need to involve the business community more and not leave all initiatives and investment to local government. The historic core is self-evidently attractive, but it needs better interpretation and access to historic and architectural sights. The city tour guide system is proving successful in this regard, but so-called cultural visits do not have to be restricted to 'museums and monuments' and should be complemented by some of the other activities mentioned. While concentrating on the cultural heart of the city – *'una identidad singular y original'* – there appears to be a need to develop and market a 'brand' for the city that is *'una marca definida, solida, consistante y diferencial de otros destinos'* ('a distinctive brand – firm, consistent and different from other destinations'). Palma is simply not well-known enough to its potential market, a feature of many other provincial cities. It has to be distinguishable from likely competitors not only because of its own place uniqueness but also in terms of access, costs and 'value for money' (INESTUR, 2008: 112). Urban tourists who include Palma in their itineraries mostly access information via the Internet, and many of the cultural resources are not described sufficiently in various languages, especially English, a lesson that many of the new boutique hotels created from old town houses in the historic core have already learned. A survey in 2007 revealed that only about 17% of British and German tourists destined for Palma used the package holiday method, the vast majority making independent arrangements for flight and accommodation via the Internet (INESTUR, 2008: 57).

Rural Tourism and the Interior

The geography of tourism of Mallorca quickly reveals, unsurprisingly, a massive concentration on the island's coasts, with the exception of much of the north-west coast where the cordillera of the Tramuntana plunges directly into the Mediterranean. One of the socio-spatial processes that assisted the growth of the coastal fringe and its urbanisation was the out-migration of people from the heart of the island to provide, in part, the workforce of the hotels, bars and restaurants and a whole galaxy of supporting activities. One consequence of this movement, inevitably, was the loss of a potential rural, agricultural labour force, especially of its younger element. In the 1950s, agriculture employed 45% of the workforce; today (2010), it employs less than 3%. Only about 50% of Mallorca's surface is suitable for farming, but since the advent of mass tourism as the principal part of the economy its frontier has been rapidly retreating (see Table 6.1 for land-use changes).

Once agriculture began to decline, one of the main raison d'etres of the smaller inland towns and villages also began to disappear. Similarly, the smaller isolated farms became increasingly abandoned. This meant that for various reasons the socio-economic infrastructure of the interior became concentrated in the larger towns that were able to support economies of scale in areas such as education and health services. Towns such as Manacor, Inca, Pollença and Felanitx, although initially affected by the centrifugal processes described above, were able to recover and expand their central place functions, aided by various planning measures.

This emptying of the countryside especially in Es Pla has presented, then, both a challenge and an opportunity to Mallorca's development in the late 20th and early 21st centuries. Falling land values from the 1960s onwards and the steady supply of farm and village houses created a resource for the development of an alternative segment of the tourism industry: rural tourism. But this supply had to be matched by a rising demand for such holidays. This was unlikely to come from the increasing number of Spanish visitors, but it was to prove very attractive to the two main North European countries familiar with the island, the Germans and the British. Both had a predilection for the countryside with a strong historical affinity for outdoor activities, recreation and relaxation that dated back to at least the Romantic era of the 19th century. To the newly affluent Mallorcans, the countryside represented the place many of them had recently 'escaped' from although some, in the early years, still practiced seasonal return migrations to old family homes. But for the former rural poor, the attractions of Palma and its suburbs and the new coastal towns far outweighed any sentimental attachment to the smallholding, the farm or the village. Into this growing vacuum at the heart of the island and in the northern and western mountain municipalities stepped the British, German and other European investors. As early as the 1950s, for example, members of the declining British colonial and diplomatic services were buying up fincas and farmhouses. Property was cheap and 'help' with household duties and gardening could be obtained for a few pesetas a day. Retirement pensions went a long way (Beckett, 1947). These foreign frontiersmen of the rehabilitation of the Mallorcan countryside were the forerunners of a wave of inward investment that saw the larger, former *possessions* as opportunities for conversion to rural hotels and centres of agrotourism (Figure 9.1).

This process was encouraged by government policy motivated by a range of tourism-based factors. To begin with, there was little concern for the drift from the countryside and the rundown of agriculture, but by the early 1990s, the pressures brought by mass tourism had become acute. These pressures were spatial (concentration on the coast), environmental (the carrying capacity of the coastal urbanisations) and seasonal (July and August) and so the potential of rural tourism as a partial alternative became better appreciated. Rural tourism was rarely seen as a counter-attraction to the coast per se but more as an additional quantitative segment to be exploited in the off-peak seasons: additional numbers, in different places, at different times of the year. It was perceived as an opportunity to spread the benefits and know-how of a well-established and successful industry (tourism) to new areas and sectors.

However, despite the growth of second homes, new immigrant settlement from the 1950s (especially British initially) and the gradual rise of commuting from the countryside to Palma and its surrounding poligonos, rural tourism really dates only from the 1990s, as Table 9.2 demonstrates.

In Mallorca the three categories in this table are used to describe the accommodation aspect of rural tourism; other CCAAs used broader, additional categories. While having one of the largest concentrations of tourism overall in Spain, in this sector the island is a tiny contributor to rural tourism nationally, providing only 2% of the total (CAEB, 2002) and less than 1% of Mallorca's tourists; numbers are reckoned to be in the order of 90,000–95,000 per annum. Such visitors tend to be middle aged and middle class – 80% are more than 30 years old, most are well educated (50% are graduates) and work in management and the liberal professions. It would appear from surveys that the majority are on repeat visits to Mallorca, suggesting that previous visits, perhaps on a different kind of vacation, are important in decision making (CAEB, 2002: 25). Accommodation in rural areas tends to be relatively expensive, and it is this that perhaps best defines its social characteristics.

Individual expenditure for this accommodation and on associated aspects of such holidays may be high, but the contribution to the tourist economy is really very small because of the small numbers involved. The number of jobs created is also small. Ownership of many of the agrotourist hotels in the majority of cases is in the hands of immigrants, especially Germans, and observation suggests that a sizeable part of the workforce is also immigrant, often from former communist East Europe. There may be other benefits to the countryside such as the restoration and conservation of *possessions* and other historic buildings, helping to restore some social and economic life to the interior especially in the low

Table 9.2 Rural tourism

Year	Rural hotels		Agrotourism		Interior tourism	
	Establishments	Beds	Establishments	Beds	Establishments	Beds
1994	3	76	42	398	—	—
2000	12	418	83	959	11	123
2006	25	887	142	1920	43	530

Source: Dades Informatives (various dates)

Photo 9.1 An island rural hotel converted from an 18C finca in Ruberts

seasons, but economically rural tourism, as defined, appears to bring little gain to Mallorca (Photo 9.1). Clearly, for the well-heeled rural tourists, the accommodation and milieu that hotels and converted farms offer is excellent, but it is producing part of that exclusive landscape that is also populated by the new permanent settlers, the second homes of affluent incomers and well-off natives and golfers. In Chapter 2, the geographer Climent Picornell drew our attention to the dangers of, in effect, 'gentrifying' the countryside or turning Es Pla into a theme park (Picornell & Picornell, 2008). Is a gulag of rural resorts isolating itself from the mainstream of Mallorcan cultural life what is really required? If the rural areas become 'touristed' landscapes will they be more about myth than reality (Cartier, 2005: 15)?

Sailing

It is hardly surprising that an island environment such as the Balearics has attracted the yachting fraternity. While certain locations in Mallorca offer large and secure harbour and berthing facilities such as the Bay of Palma, much of the rest of the coast does not possess good natural facilities. Much of the north, south and east coasts have shallow, naturally shelving marine structures. Historically, these could be utilised only by beaching vessels from these shallow waters; even Palma used this system below its city walls until the first mole was built in the 14th century. The west coast has few accessible harbours, the principal exception being Port de Sóller. By 2008, there were more than 14,000 sailing berths in Mallorca, attracting nearly 300,000 'sailors', over half of whom actually fly out to their boats (*Dades Informatives*, 2008: 104–105; INESTUR-CAEB, 2007).

During the period up to the Civil War and the Second World War, recreational sailing remained the preserve of the monied few, but a remarkable transformation has taken place since the 1970s, with sailing increasingly becoming a middle-class sport or pastime. While some of the world's largest and most expensive cruisers can now be found in Palma's marina, the majority of sailing vessels are of more modest size. For local people, there has been a revival of the *llaud*, a simple vessel used for recreation at weekends. However, the yacht or cruiser has now, in effect, become an addition to the accommodation structure of Mallorca's coastal towns. Theoretically, temporary or seasonal larger vessels have now become floating apartments, with many owners rarely straying from the safe anchorages. To encourage and accommodate this new addition to the diversification of tourism, many local authorities in Mallorca have constructed massive breakwaters, moles and havens from huge boulders deposited in the shallow waters to provide deeper berths. These developments have met with mixed responses from local people. The municipalities and the *clubs nauticos* see them as new income streams from yachtsmen and women, adding to tourism numbers and in some cases extending the tourist season. In recent years, berthing fees have risen considerably, usually in advance of inflation rates because of the belief that boat owners, who are predominantly foreign, can afford them. Those opposing the continued expansion of marinas in established and new locations see them as contributing little to the local economy (perhaps one reason for the increase in fees), polluting the marine environment and helping to destroy the stands of *Posidonia oceanica* that help protect natural beaches.

In morphological terms, many of these marinas show very dense occupation and sometimes detract from the front-line status of expensive hotels. They appear as extensions of the urban structures to which they are attached. A good example is the creation of the marina between Cala d'Or and Porto Petro on the east coast where the mixing of buildings and boats makes up one complex urban landscape with nearby car parking and a proliferation of shops, restaurants and marine services. As with other aspects of urbanisation, the planning control of marinas has lagged behind actual construction.

Spatially, there are marked variations in geographical distribution. Figure 9.3 shows that the greatest concentration of berths occurs in almost exactly the same localities as the majority of land-based tourism – Calvià and the Bay of Palma. Their location is not determined by natural

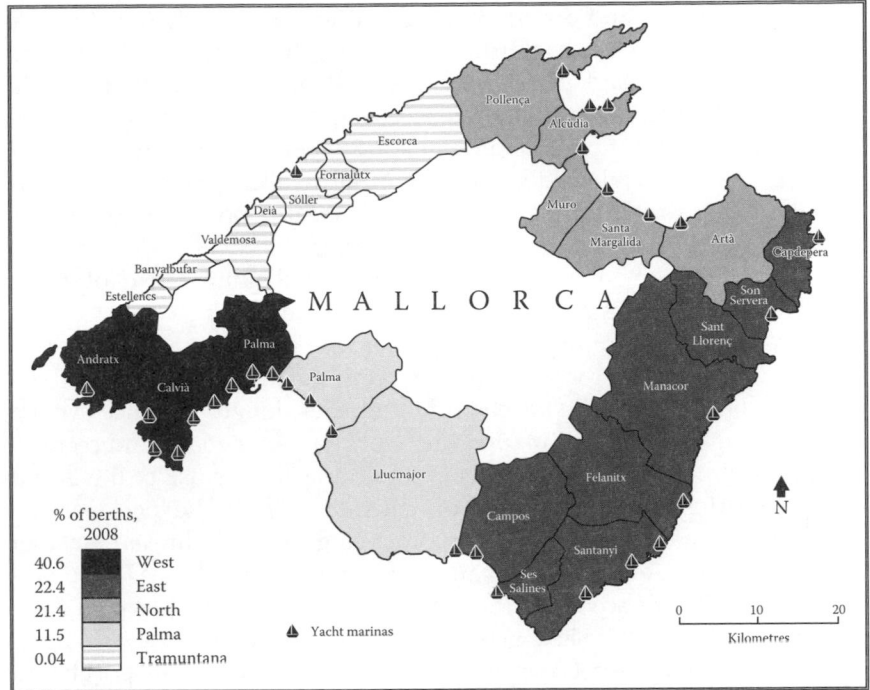

Figure 9.3 Map of sailing marinas. *Source: Dades* Informatives (2008) and own calculations

harbour facilities but, as we have shown, by artificial marina construction in the most aggressive municipalities, further increasing the spatial pattern of tourist activity.

Hiking, Walking and Cycling

Data on these forms of outdoor leisure activity enjoyed by visitors and local people alike are hard to find. Personal observation suggests that there has been a rapid increase in the number of hikers and in investment to service their needs. Some of this investment came from the short-lived ecotax in the early 1990s. Given the weather conditions, it is not surprising that the high season tends to be shunned by many walkers and the months of December, January and February can be wet with quite intense downpours, leaving the main season for walking to be in the spring and early autumn. The area most popular is the Tramuntana where a network of paths has been inherited from all eras from the Arab development of the mountains to medieval charcoal burners and snow and ice collectors to 20th-century tobacco smugglers. Some of these are beautifully stepped and paved, others forming part of Ludwig Salvator's late 19th-century estates (Ordinas Garau, 2000). Today, many are waymarked paths laid out by the Consell and local authorities. In addition to the path and trails, there is a set of *refugios* for those intent on longer distance and more strenuous hiking. There is now a much better collection of maps and guidebooks to support walking in the mountains (Alpina maps, various dates; Dillon, 2009). While popular with all ages and nationalities, fieldwork observations suggest that youth and the more elderly German and British visitors make most use of these resources. For the less strenuously minded, many of the coastal tourist information offices supply walking guides and maps to the local area.

Cycling takes three forms within tourism: the hire of bikes on a daily basis for short trips to the beach and other sites, longer-distance cycling holidays using a now well-mapped network of minor roads and country lanes (*camis*) from traditional hotel bases and lastly, the use of the island by cycle touring clubs, the *peletons* of which range from the local German club to the professional racing team practicing for the Tour de France or the Giro d'Italia. An attractive feature of Mallorca for all these groups is the wide variety of scenery – and degrees of difficulty – derived from Mallorca's diverse physical geography. The construction of the light-weight modern high-speed or mountain bike means that transport by air is no longer a problem.

Has an Alternative to Diversification Been Found?

Despite considerable effort on the part of Mallorcan hoteliers and resort authorities, the attempts at diversification have had limited success and have to be viewed relative to the continuing increases in the traditional sea, sand and sun model. An overall increase in the total number of visitors whether 'diversified' or not means that externality costs have increased (Williams & Shaw, 1998: 55). In spatial terms, it is interesting to note that many of the diversified forms described are geographically limited to Greater Palma and the south coast area. Even away from that zone, they appear almost as fringe-belt activities to coastal urbanisations, simply extending the landscape of coastal mass tourism further inland.

'Diversification' need not be confined to the types of alternative holidaymaking described here. It has often been the case that much of the actual diversification has occurred within the basic product of mass tourism in Mallorca, the package holiday. Much effort has been directed towards meeting the changing consumer demand within this proven system. As many authors have pointed out, it was once the objective of travel and tourism to discover and enjoy the benefits of the exotic, the different, but as John Walton has written: tourism '... feeds on local differences, puts the experience on sale, and through commodification converts it into artifice, transmuting the objects of the gaze by making them aware of the monetary value of an authenticity which, as a result, becomes illusory' (Walton, 2002: 123). The exotic has increasingly become more and more eschewed over the last two decades. Instead, there have been increased efforts to render the Mallorcan scene more and more like that 'at home'. This is by no means a new phenomenon; writers in the 1950s noted the introduction of many signifiers that had direct appeal to the tourists' native culture. There are many symbols of this ranging from the considerable increase in car hire to restaurant menus. The production of tourist space has in many ways become more limited in content and more circumscribed geographically. Today, in many of the more popular resorts, nationality symbols are all-pervasive and so parts of many of them are to all intents and purposes ghettos of British or German popular culture. The food, the TV programmes, the songs and music, the bars, clubs and discos, even the shops names are all too familiar, often complemented by a regional or urban social solidarity for Geordies, Scousers, Valleys girls and Düsseldorfers (Photos 9.2–9.6). This spatial reinforcement in places such as Paguera, Magaluf, Santa Ponça and Palmanova means that many tourists rarely travel outside their resort's main streets, a trend reinforced by the move to all-inclusive holidays that

Photo 9.2 British-run café in peripheral location in Magaluf

Photo 9.3 Magaluf: Just like home

Photo 9.4 Magaluf: Pandering to the lowest common denominator

Photo 9.5 Magaluf: Food from any British High Street

Photo 9.6 German stube in Platje de Palma

financially confine visitors to their hotel and its grounds (Andrews, 2006: 224–227). In addition, in many resorts some tourists seek to take to extreme some of the trends in youth culture practiced at home especially drunkenness ('binge drinking') and sexual liberty. A survey conducted in 2007 of 3003 German, English and Spanish 16- to 35-year-old tourists in Mallorca and Ibiza (1476 in Mallorca) purported to show that more than half of the respondents had chosen these islands for their nightlife: 60% admitted having got drunk more than two days per week and 6.4% said that they had been involved in a fight. Illegal drug use was also prevalent. The most important factors in rating night-time venues were cheap drink (78%), opportunities for sex (63%) and loud music (62%) (Hughes *et al.*, 2008). On a somewhat different scale, this is not a new phenomenon. In the 1970s, many tourists were attracted to 'Majorca's' cheap alcohol and opportunities for sexual indulgence; Akhtar and Humphries record many examples of excesses at that time (2000: 120–123). Such behaviours are hardly exotic nor are they enacted in exotic locations. Is it possible that the current trend of binge drinking in Great Britain has its origins in holidays in 'Majorca'?

Can the continuing success of the mass tourists' package holiday – which has confounded many observers and led to many Mallorcan intellectuals' frustration – be ascribed to this kind of innovation of cultural reinforcement? How much of this cultural repetition is ascribable to Mallorcan ingenuity in reacting to consumer demand? Or is it more the product of British and German investment made possible by the local rental systems for everything from property to even knives and forks that reduces risk but abrogates responsibility? While the hotels of the large chains remain largely immune from this trend, the penumbra of low-grade urban activities that surround them appears to be increasingly in the hands of small-scale overseas entrepreneurs. Any visitor to Magaluf – that quintessential British resort, although German numbers are increasing – is only too aware of this on the streets at the heart of the resort. Large-scale Mallorcan capital has long since found an outlet in overseas investment, particularly in the Caribbean and Latin America, as was demonstrated in Chapter 8.

Perhaps the real enemy to diversification is the all-inclusive holiday pioneered by Club Med but now practiced by practically all the integrated hotel chains and marketed in conjunction with British and German tour operators (Furlough, 1993, 2009). Allegedly giving financial benefits to the tourist and the hotel, their impact on local economies has been profound with the use of bars, cafes and restaurants declining (Aguiló *et al.*, 2003). More recent statistics, however, show the number of bars, cafes and restaurants increasing from a low point in the early 2000s up to 2008 (*Dades Informatives*, 2008: 98–102). The current slowdown in the economy will inevitably affect this rather fragile sector.

Ten years ago, there were apparent national differences in the take-up of such holidays (Kozak, 2001), but today they are much more ubiquitous and appear to give a high level of satisfaction (Llibre Blanc, 2009: 152ff). All-inclusive holidays are a convergent activity, leading to economic, cultural and spatial concentration that will surely further exacerbate Mallorca's attempts at product diversification.

Chapter 10
Future Trends

Introduction: Adjustment or Seismic Shift?

As Mallorca enters the second decade of the 21st century, its tourism industry may not be facing a crisis but it is certainly having to adjust to changing circumstances. These now include the world recession, which began in late 2007 and continues as most countries – and many regions – strive to find a way forward. Although Spain's banking and financial services sector was initially thought to be less affected than those of the United Kingdom and the United States, other sectors of its economy have been devastated, notably its housing and construction industries. Mallorca with its extreme dependence on a broadly defined tourism industry, which now includes residential tourism, remains anxious about its future. From its beginnings in the early 20th century, and especially since the advent of mass, low-cost tourism in the 1960s, the island's economy has been inextricably linked to the fortunes of countries to the north and west of it that have sent it holidaymakers and visitors in their millions, millions whose disposable incomes for holidays have depended upon economies over which Mallorca has had no control. On two occasions before the recent recession (post-2007) – in 1973–1975 and 1991–1992 – the ever-upward curve of visitor numbers trembled but recovered. Will the same kind of elastic recovery occur this time around after the worst economic crisis the world has seen since the 1930s? Has the tourism economy already made enough adjustments from its peak in the 1980s to cope with a prospective downturn in numbers and income? Have the diversification strategies of the last 20 years, one of which was to reduce environmental impact, been sufficient to attract new genera-tions of tourists? By May 2010, overseas visitors to Mallorca were down by 12.5% when compared with the previous 12 months, with only the rural and cruise sectors showing any growth. The number of British visitors was down by 16.6% (Document 207, May 2010, www.inestur.es/documentos/859_mi.pdf). Is Mallorca at last entering the terra incognita of the stagnation and decline sections of Richard Butler's TALC (Tourism Area Life Cycle) or is this recession likely to prove to be simply another shudder in the upward trend of the curve?

In 2009, a major review of tourism was undertaken to try to define future directions. This was timely because the number of visitors to Mallorca fell by more than 1 million from 2008 to 2009, causing some concern. In some ways, this was similar to the exercise undertaken in 1986 in that university academics and the Govern Balear cooperated to produce *Llibres Blancs* (White Books). The latest version succinctly expressed the view that 'The Balearic tourist system finds itself at a crossroads, in a phase of transition, between what it was and what it will be. Some of its comparative advantages are on the brink of disappearing, without there being any new ones in sight' (Llibre Blanc, 2009: 360). Preceded by very detailed analyses of tourism in the Balearic Islands, including an examination of the changing dynamics of its economic structure, its environmental, spatial and demographic patterns and its present position in relation to overseas competition, it concludes with a section on the challenges of the future (Llibre Blanc, 2009: Part 5). Its perspective on the future is largely derived from an academic analysis of what is required to ensure the future success of the industry. This might be summarised as seeing a need for cooperation between the public and private sectors within an institutional framework of 'planning' in the broader sense. Four approaches are stressed: the need to recognise and act upon environmental and resource constraints ('sustainability'), the need to maximise social welfare benefits for the citizenry while permitting profit satisfaction for the private sector, the need to innovate and lastly, to ensure that all actors recognise the growing competition from other destinations worldwide ('competitivity'). The kinds of adaptation to changing circumstances that have characterised and benefited tourism in Mallorca in the past will no longer suffice according to the authors; the entrepreneurial spirit identified here in Chapter 8 will need to be directed more towards cooperation between the private and public sectors (Llibre Blanc, 2009: 371). Or put another way, Mallorca's tourism future is unlikely to be derived from market solutions alone. This clearly has a strong ideological overtone, a plea for much more intervention by government, especially by the Govern Balear. While wholly admirable in its sentiments, this will be dependent upon the political structures of future governing elements in the *Parlement* and the political strength and will of the increasingly powerful multinational hotel, airline and tour operator corporations who manage so much of the island's tourism industry, factors examined here in a previous chapter.

At this point in time, it remains clear that for the proximate future Mallorca will have to continue with much the same form of tourism in order to drive its economy and maintain its residents' standards of living. Despite the movement into new forms of employment in the tertiary and quaternary sectors – and to a lesser extent in manufacturing – despite the early promises of the construction industry, despite the regeneration of the labour and housing markets through immigration and demographic growth, despite the rise of indigenous consumerism and very conspicuous consumption by islanders, Mallorca remains – and will remain – highly dependent upon tourism and within that sector, upon tourism en masse.

On the reasonable assumption that the recession of 2007–2010 will eventually recede and more normal trading resume, the growth in number of visitors will go on rising in the medium term. Although visitor numbers to Mallorca fell 11.2% in 2008–2009 and advance bookings for 2010 initially appeared to show an upward trend continuing (LIRC, 2009: 61), the reality of actual bookings fell considerably short of expectations. For the near future, it would seem that Mallorcan entrepreneurs remain confident of their faith in the pre-existing 'sun, sea and sand' model of tourism that has made their fortunes over the last 70 years. The elasticity of demand from consumers within the Balearic Model – despite its revision by forces of globalisation – should ensure a continuing flow of tourists from Europe, possibly supplemented by numbers from Asia.

History of the Balearic Model

From the late 1960s, most forward planning for tourism in Mallorca was based primarily on extrapolation of previous data. While this was fairly satisfactory in the early days, the energy crisis of the early 1970s showed its vulnerability; for the first time, a levelling out of the upward curve made predictions less certain. This was repeated in the early 1990s when the German economy slowed down following the reunification of the country and again in the world recession beginning in 2007. Mallorcan hoteliers and government planners began to realise that the Mallorcan tourist economy was no more self-contained than its agricultural or industrial economies. Its future would increasingly be dependent upon the performance of other countries. The island also had to accept that its pioneering work in the development of the modern mass tourism market was now over, with other parts of Spain and many other areas of the Mediterranean offering competition for its once assured markets. French investment in its own Mediterranean coastal settlements west of the Rhone had already created intervening opportunities for French tourists originally destined for the Costa Brava as early as the 1970s; the Greek Islands were opening up once jet plane travel and island airports reduced travelling time from the United Kingdom and Germany. By the 1990s, the Turkish coast could offer package holidays at lower or equal costs to those offered by the Balearic Islands. This competition from the geographical expansion of Mediterranean destinations was in itself not a problem so long as three conditions could be fulfilled: the continuing growth of the European leisure market via social and economic advancement (i.e. more disposable income and increased leisure or non-work time), secondly, demographic growth in the despatching countries and, thirdly, whether Mallorca's tourism structures were robust enough to respond in order to retain their market share.

The growth of the Mediterranean and indeed worldwide market from the late 1980s has been covered elsewhere (e.g. Manera *et al.*, 2009: 1–10). For our purposes, it is important to focus on events in Mallorca, particularly the growth of the German market, which overtook British tourists to the island in 1996. This was not really substitution but rather additionality, but it did demonstrate that there were other tourist markets that could be tapped. The same process could be, and was, identified for other countries notably from Peninsula Spain itself and from the newly emerging countries of Eastern Europe, a process we described in Chapter 4. The assumption here for planners was that the traditional model could simply be expanded to accommodate new numbers with the same kind of holiday packages that had always proved successful – the Balearic Model for consumers. Although a new flexibility and response was required, by and large it was a continuation of previous patterns. We have seen the continuing dominance of the three-star hotel, the emphasis on low-cost beach locations and a tolerant planning system. There were new aspects to this process. Apartments and self-catering were added to the accommodation register, the airport at Palma was massively expanded as were other transport systems, many older hotels were refurbished and many beaches artificially revived. While environmental issues were rising up the political agenda resources for tourism continued to be made available although at rising costs.

However, policymakers wished to follow an additional strategy, that of diversification (Chapter 8). The broadening of the tourist year to include mid and low seasons and the introduction or expansion of new forms of holidaymaking, including golf, sailing, cycling and hiking, certainly added new dimensions to tourism, but their effect was to add additional numbers, not redistribute existing ones. The same effect could be identified for the expanding cruise and conference markets. The pressure on resources was not ameliorated but exacerbated.

In what sense, then, might Mallorca's tourism industry's responses to changing forces be described as innovative? The notion of R&D leading to invention and innovation is a relatively new aspect of tourism management. In manufacturing industry, R&D leads to new products and new processes for producing old products at lower cost. In the tertiary and quaternary sectors, the same objectives apply but the processes are more difficult to identify and describe. Many of the new features of the tourism industry have in reality been transfers of technology from other sectors. In the hotel industry, electrical and electronic improvements have transformed such things as mass catering, heating, lighting and building services. In the transport sector, the jet plane and the no-frills airline revolutionised access to Mallorca. In the automobile sector, the concept of hired cars radically altered personal mobility while on holiday, helping to disperse tourist pressures away from the beach resorts as has the side tour by coach. New forms of activity, such as safari parks, water parks, coastline cruises, glass bottom boats and historic tours to towns and gardens, have been introduced, but these changes have simply been propelled by the rising demand, qualitatively as well as quantitatively (Photo 10.1). In the energy sector, the constraints felt from the 1970s onward by Mallorca's generating capacity have been met by new build, raising output and efficiency, building new transmission systems and above all, and more recently, by connecting to mainland supplies by gas pipeline and undersea cable. There are long-term costs associated with this, of course, but the higher per capita incomes of Mallorcans and their tourists can surely afford it. If supplies from the Peninsula can be brought in for electricity and gas, why not water supplies too? Is the technology very different? It may prove to be a more feasible solution than continuing investment in expensive desalination plants. European Union cohesion funds may be available too. Technology can change some of the resource balances within Mallorca.

This raises the question for tourism planners in Mallorca as to whether or not the tipping point in the Butler model has been reached: can the year-on-year increases in tourist numbers continue – and it certainly looks inexorable, despite the decline in 2008–2009? If not, what factors might constrain its upward rise or even lead to a long-term decline?

Photo 10.1 Aquarium Platje de Palma: A distraction from the beach

Traditional theory would suggest that the answer might be sought in the ecological notion of carrying capacity, a concept examined in Chapter 6. This is essentially a theory based on a particular understanding of natural ecosystems whose transfer to human ecology is more debatable. Instead, critics of the continuing growth of tourism turned to environmental degradation and the limits of resources to support this kind of economic activity. For islands such as Mallorca, any model of expansion is constrained by its physical limits (Mayol & Machado, 1992). However, it is clear that at this point in time these limits have not yet been reached. There are a multitude of reasons for this amongst which we can identify five elements: the changing nature of the tourist experience including the partial shift from hotels to apartments and towards residential tourism and a secular move away from beach-based holidays; the reduction in seasonality; technological advances to improve the position of the supply of goods and services and to ameliorate tourism's impact on resources and environment; the move of part of the economy and its associated social infrastructure from tourism to other forms of activity aided by the growth in the labour force through immigration and lastly, the effects of globalisation on the island's economy and people. The Llibre Blanc pointed out that in theoretical terms,

> ... economic activity in general and tourism in particular, more than a process of transformation is a process of the creation of value... In fact, economic growth is no more than an increase in the flow of generated economic value. Therefore, although tourist growth is certainly limited from a quantitative viewpoint, growth measured in terms of value does not have to be so; whenever feasible, through transformation, it can create more value, more utility, from the same available quantity of material and energy. (Llibre Blanc, 2009: 376)

All this points to the continuing growth of tourism and the island's capacity to manage it. This is not to advocate of a particular policy for the future; in many ways, it simply exemplifies what has been happening over the last 50 years, a direction that would be difficult to change in the near future.

The crucial political issue for Mallorca is whether or not it wants to continue along this path set out for it by decisions made many years ago, even those from as far back as Amengual's arguments to support tourism from the late 19th century. Nearly all of the local arguments against expansion of tourism – social, economic, environmental – have been couched in terms of retaining the historic Mallorca, its landscape, its rurality, its language, even its history, research into which, incidentally, has been dominated by a concern for the brief medieval kingdom of Mallorca rather than the more relevant last 150 years, all of which are elements familiar to students of European nationalism. No one in

Mallorca would or could advocate the removal of tourism from Mallorca, but there has always been a strong section of local society that has argued for either slowing down the rate of growth or changing its nature, especially supporting the idea of moving a significant proportion of it 'up market' i.e. a structural change involving fewer numbers but maintaining at least current income from tourism.

This latter point might be couched in sociological or – more bluntly – class terms. Would Mallorcans wish to see a marked reduction in mass tourism, which by the definitions used here means primarily low-income tourism, even working class tourism, followed by an increase in bourgeois tourism? Certain benefits might be gained from this, especially in terms of resource use and environmental impact, but does the island have the right resources and infrastructures in sufficient quantity and of sufficient quality to support such a move? The same question could reasonably be asked of the tour operators supplying such tourists. For both sets, inertia is a powerful force, difficult to shift. Although new tourist products are beginning to be established on an increasing scale, much more investment would be required (Salvà Tomàs, 2000: 128–133). In any case, many of these new forms of (bourgeois) holidaymaking – golf, marinas, second homes etc. – are already the subject of considerable opposition from certain sections of Mallorcan society. We have pointed out before that so far most of these diversified forms of activity have largely been aimed at new and additional markets: tourism and its impact continues to grow but in new guises. Such arguments are by no means restricted to Mallorca; they are encountered all across the Mediterranean; see Spilanis and Vayanni (2004: 269–291) on the Aegean Islands for example.

An important weakness of the Llibre Blanc's 2009 prognostications for tourism is the apparent neglect of the importance of international tour operators from their analysis. In the late 1990s, more than 90% of arrivals to the Balearic Islands were via tour operators; only 8% were independent. More than half of all package holidays at that time were controlled by the five largest tour operators, with another 48% shared between a myriad of smaller tour operators (Sastre & Benito, 2001: 78). These data have changed in two respects over the last decade: the number of tour operators has been reduced by merger, acquisition and takeover, putting them in a stronger, almost oligopolistic bargaining position, and independent travellers using the internet and no-frills airlines have increased. In Chapter 8, it was noted that while large numbers of holidays may be initially identified and booked via the internet, they are routed through the websites of the major tour operators who now encourage such methods to reduce costs. Indeed, it is the case that this kind of innovation has been more quickly and deeply adopted by larger tourist organisations than by smaller ones since managerial attitudes are more receptive. Government policy in this field will need to be focussed on the smaller hotel and transport agent where technical training and help will be required for website construction and use (Garau & Orfila-Sintes, 2008: 78). It is in the non-mass markets that the Internet is used most for bookings in areas such as golfing holidays and agrotourism. Tour operators still command a significant part of the Mallorcan market and they, on behalf of the prospective customers, will be constantly searching for new locations and value-for-money locations within the Mediterranean and elsewhere. Any negotiations between the private sector and government to enhance welfare benefits such as the Llibre Blanc advocates must include these multinational corporations. In the histories of tourism in Mallorca by Mallorcans, the role of British and German tour operators is being sadly underestimated. Mass tourism in the past and its continuation into the medium term must include these important actors. We have drawn attention to inertia above, but fickleness and changing tastes can in time alter patterns of demand. The dependence of Mallorca on tour operators is increasingly paralleled by a similar role for airlines. Self-managed bookings via the Internet and the use of low-cost airlines such as Air Berlin, easyJet and Ryanair will also mean that these players must also be brought into any negotiations on future plans for tourism in Mallorca. Recent events have exposed the fragility of some of these airlines, which may well lead to more mergers and a stronger negotiating position for larger firms.

Some movement can be detected, then, towards more individualism amongst potential visitors to Mallorca using the Internet and no-frills airlines or seats-only bookings with tour operators such as Thomson. This is unlikely to undermine the dominance of the major tour operators for a destination such as Mallorca in high season, but there is growing evidence that off-peak holidays in hotels and even high-peak holidays in golf resorts and in agrotourism are benefiting from this trend (Vich i Martorell & Pou, 2007). Since about 2000, some consumers, thanks to these innovations and the related cost factors and to the new flexibility in the British labour markets, have begun to spread holidays over wider periods of the year, moving away from the traditional sun and sand holidays towards more culturally diverse activities. This rise in 'individualism' and movement against being part of 'mass' holidays, often related to the rise in environmental concerns, especially amongst German tourists, have all helped reduce the traditional package holidays and the large influence of the corporations that supply them. Salvà criticises so much of today's tourism in Mallorca as the neofordist creation of the hotel and tour operator industries, characterised by its indifference to the island's environment, history and culture. In his words, highly urbanised resorts have become 'tourist concentration camps' – an unfortunate phrase also repeated by Cirer (2009: 324), that are self-contained worlds, part of a no-place realm or worse still, part of mythological, theme park geographies (Salvà Tomàs, 2000: 178). By the beginning of this century, there were 13 tourists for each resident; it is said that they had little or no interest in the culture of the area. The new European residents were similarly isolated from the locale; they saw the island as a refuge and thanks to new technologies such as the internet and satellite TV also had no use for local culture (Seguí Pons & Martinez Reynés, 2002: 213). If the 'ghettos' of mass tourism, compounded by all-inclusive packages, were not to be repeated by the chosen isolation of the new residents, ways would have to be found to integrate them closely into Mallorcan society. For the future, Salvà points to 'the rise of a flexible, staggered and individualized leisure industry' with 'greater quality, authenticity and personalization of services' organised around the four Es: environment, equipment, evenness and *enquadrement* (within a framework) (Salvà Tomàs, 2000:177). Similar arguments have been put forward more recently by Carles Manera, currently the Director of the Ministry of Economy and Finance for the Govern Balear but also Professor of Economics in UIB (Universitat de les Illes Balears) (Manera, 2010). Their analyses and proposals are well made, but both would appear to underestimate the extent to which the tourism economy of the Balearic Islands is actually controlled by outside forces and agencies: the probable patterns of North European tourists' demands in the near future and their satisfaction being met by multinational tour operators working with increasingly interna-tionalised Mallorcan hotel chains.

Thus, it is necessary to reflect again on what might be the role of tourists' demand rather than the features of Mallorca's supply in the future evolution of the island's economy and society. Mallorcan hotel owners, resort developers, planners and politicians have created a geography based almost entirely on demand, whether that be in the form of a mass tourism complex on the south coast or a golf course inland. Examination of the history of the hopes and aspirations of the early pioneers of the industry such as those who built the middle-class resorts based on garden city principles such as Cala d'Or or the luxury hotel of Adan Diehl and even the modernist hotels of Luis Gutierrez Soto has been drowned out by the demands of the tourist en masse. The early principles laid down in a more gentle age of luxury hotels, cruise liners and garden suburbs soon gave way to a landscape of mass tourism from which both sides have reaped enormous benefits: cheap and initially exotic Mediterranean holidays for north Europe's working classes, and jobs, wealth and social security for Mallorca's people together with considerable profits for her capitalist class and overseas tour operators and airlines. Almost from the beginnings of mass tourism, many individuals and groups began to protest at the direction tourism was taking and the effects it was having on the island's society and environment; there were fewer protests about its economic benefits except in terms of its dominance. Many of the early objections came from what might be termed the indigenous population of Mallorca, that is, the population before the great migratory movements of workers into the island in the 1960s and 2000s. Twenty years ago, Climent Picornell (1990) listed 18 general conclusions drawn from the effects of tourism on Mallorca, nearly all of them negative.

Mallorca was far from being some kind of 'virgin' island before the mid-20th century. Changes had occurred in the island's economy many times – some of these were documented in Chapter 2 – supported by the arguments of Manera and others who saw Mallorca as more in the mainstream of European culture from the 18th century onwards than many would previously accept. However, the kinds of changes and their rate were largely under the control of Mallorcan society and took note of, and built upon, Mallorcan culture and history. The introduction of mass tourism, notwithstanding the preceding forms of tourism in the 1920s and 1930s, was essentially an alien intrusion. The outside forces of overseas agents of change especially British, and later German, travel agents, tour operators, hotel companies and transport organisations had little knowledge of, and perhaps less sympathy for, Mallorcan culture. This coupled with local developers' aspirations, and avarice, transformed the Mallorcan landscape, economy and society. Visitors en masse have brought with them their own geographies and things to 'gaze' upon: Magaluf is not the product of Mallorcan culture but that of any British city or town on a Saturday night (Photo 10.2). The appearance of

Photo 10.2 Magaluf: Multiculturalism in food

the bars and restaurants, the language of commerce, the signage, the food, the music and dancing, the large screen TVs showing Premier League football and the sexual depredation are all translations from Manchester, Newcastle, east London and a hundred British towns in between (see Photos 9.2–9.5, 10.3 and 10.4). Likewise for Peguera, with its imposed German popular culture. In an earlier analysis, this author repeated the oft-quoted aphorism that many tourists to Mallorca are not even aware that they are on an island: 'they bring their geographies with them; the Balearics merely add sunshine and a different currency' (Buswell, 1996: 317).

Pere Salvà's observations of the future direction of tourism in Mallorca focus on the apparent rise of the 'new' tourism, a view that is primarily determined by new tourism products, that is, by the supply side of the equation: 'build it and they will come'. His principal recommendations for future development focus on new paradigms for tourism that include, inter alia, a greater emphasis on the needs of the individual rather than the group, environmental sustainability, cultural authenticity, activity holidays, quality of service and geographical and temporal spread. This is essentially a prospectus that has strong middle-class appeal. Unfortunately, the history and geography of tourism in Mallorca is much more a product of demand and Mallorca's response to this. If

Photo 10.3 Multicultural signage in Cala Millor

Photo 10.4 British signage in Cala d'Or

this process continues to hold sway, only changes in what tourists want from their resorts will alter the type of tourism provided. Salvà recognises this when he writes: 'The sun and beach tourism ... without a doubt will continue in the Balearic Islands in the next decades of the twentieth century (sic) but the difference between it and the new model will be based on "new products and new behaviour rules"'. The history of demand seems to suggest that such new products will only partially solve the problems brought about by mass tourism. In any case, innovation works equally well within the traditional mass sector, witness the growth of alternative attractions to the beach in such activities as water parks, safari parks and the upside down house Katmandu. The urban cancers that Salvà refers to are restructuring themselves to accommodate new forms of mass tourism, often an unintended outcome of well-meaning local authorities intent on improving resort infrastructures. Currently, more and more of the recreational activities now take place after dark and have little to do with the Mediterranean climate (Rullan Salamanca, 2006). At a more extreme level, sun, sea and sand tourism is giving way to sex, Smirnoff and sleep tourism for many young people. For private sector entrepreneurs, there is much less risk and the likelihood of higher returns on investment in this kind of demand-led modification to traditional models than in new forms of off-season, middle-class recreational activities such as golf and sailing. It is also a reflection of the ability of the small-scale entrepreneur who may be able to respond flexibly and rapidly. On the other hand, investment in, say, a marina or a golf course requires much greater capital with longer returns and a lot more consideration of planning regulations and environmental constraints. Much of mass tourism in Mallorca is still based on short-termism. This is reflected in part by the nature of the bar/restaurant/holiday goods retailing sector located on the sea front or nearby where practically everything is based on a rental system. Thus, many small entrepreneurs – often overseas born as well as local – are tempted into the holiday service trade with little capital requirements and easy exit if sufficient income does not materialise. There has been much criticism locally of the all-inclusive holiday, usually the province of the larger hotel chains, which increasingly is said to dominate many resorts; nonetheless, the product that they offer within the confines of the hotel and its grounds remains almost identical to that offered outside.

Conclusions

As the industry contemplates its future direction, it is clear that change will probably come slowly. Regulation and legislation will gradually ameliorate the worst excesses caused by the number of tourists and the pressures they impose on island environments and resources. Technology will gradually expand opportunity, particularly in the energy sector and in the creation of new tourist landscapes. Salvà's new tourists will demand new products and services, but if the island neglects its core business of mass tourism based primarily on accessibility and price, there is a strong likelihood that the large numbers of British and German holidaymakers will migrate elsewhere to where other regimes will meet their demands. Government and the political process may try to determine the direction of change according to its current ideologies, but tourism remains a market-driven activity and business will assess the risk inherent in any radical change.

Sources and References

The Major Locations of Material in Mallorca
(See Acknowledgements for Comments on Libraries)

Note: During 2010 some major changes in the structure of the ministries of the Govern Balear took place, and it is likely that further reorganisation that could affect tourism may take place. On 25 March 2010, the Govern Balear announced the merger of a number of governmental tourism institutions into one body, noticeably combining the strategic and research arm – INESTUR – with the marketing and promotion activity located in IBATUR to become Agència de Turisme de les Illes Balears (ATB). In addition, in May 2010, certain other tourism agencies were to be placed under the direct control of a new Ministry of Tourism, including CITTIB (Research and Technology in Tourism), IQT (Quality in Tourism) and IMET (Improving the Environment for Tourism). All this has been part of President Antich's policy of rationalising government departments in order to meet savings' objectives made necessary by national and islandwide reductions in public expenditure. More responsibility for the management of tourism will be delegated from the Govern to the island *Consells* (Ultima Hora, 23 February and 11 May 2010). As a result of these changes, access to material may no longer be at those locations listed below. Researchers are advised to contact the Conselleria de Turisme at the Calle Montenegro address below in the first instance. Most governmental websites are available in Catalan, Castilian, English, French and German.

Biblioteca Fundación Bartolomé March, Palau Reial, 18, 07001 Palma. Tel: (+34) 971 711122. www.fundacionbmarch.es
Biblioteca Lluís Alemany, Edifici de la Misericorda, Via de Roma 1, 07012 Palma. Tel: (+34) 971 219539. www.conselldemallorca.net/biblioteques
Bibliotecaria, Centre de Documentació. Agència de Turisme de les Illes Balears (ATB). Carrer Rita Levi, s/n-Parc Bit, Ctra. Valldemossa Km. 7, 07121 Palma. Tel: (+34) 971 177210. www.observatoridelturisme.caib.es
Consell de Mallorca, Consellera Executiva d'Economia i Turisme, Plaça de l'Hospital, 4 (Centre Cultural la Misericòrdia) 07012 Palma. Tel: (+34) 971 173894. www.conselldemallorca.net
Conselleria de Turisme: Calle Montenegro 5, Palma 07012. Tel: (+34) 971 176199. http://turismeitreball.caib.es
Conselleria de Turisme i Treball: Plaça de Son Castelló, 1 (Polígon de Son Castelló), 07009 Palma. Tel: (34) 971 176191/971 176300. www.treballiformacio.caib.es
Fomento del Turismo de Mallorca, C/ Constitucio 1, 07001 Palma. Tel: (+34) 971 725396. www.newsmallorca.com

IBESTAT (Statistical Institute of the Balearic Islands), C/San Sebastian 1, 07001 Palma. Tel: (+34) 971 78 45 75. www.ibestat.cat
INESTUR Edificio INESTUR 4, 07121 Palma. Tel: (+34) 971 177210. www.inestur.es
Universitat de les Illes Balears, Servei de Biblioteca i Documentació, Cra. Valldemossa, km 7.5, 07122 Palma. Tel: (+34) 971 173000. www.uib.es/ca/infosobre/serveis/generals/biblioteca
Xarxa Biblioteques Municipals (network of municipal libraries). www.consell demallorca.net/biblioteques/bibmunicip Centre de Recerca Econòmica (UIB • "SA NOSTRA")

The Principal Statistical Sources (See Note Above)

Statistical data produced by various tourism agencies in the Balearic Islands are being modified to conform to national statistical criteria. In the meantime, this can lead to some discrepancies between sources. The statistical data used in this book have largely been drawn from INESTUR's sources.

CRE (El Centre de Recerca Econòmica UIB-Sa Nostra): www.cre.uib.es
IBESTAT (Institut d'Estadística de les Illes Balers): www.ibestat.cat
INE (Instituto Nacional Estadística): www.ine.es
INESTUR *El turisme a les Illes Balears. Dades Informatives*: www.inestur.es
The Tourism Observatory – much of the online information formerly available from INESTUR is now at www.observatoridelturisme.caib.es

Bibliography

The following references are a fairly comprehensive guide to the academic literature that deals with Mallorca's tourism industry.

Abram, S. and MacLeod, D. (eds) (1997) *Tourists and Tourism: Identifying with People and Places.* Oxford: Berg.
Abulafia, D. (2002) What is the Mediterranean? In D. Abulafia (ed.) *The Mediterranean in History* (pp. 11–31). London: Thames and Hudson.
Aena (2010) At http://estadisticas/aena.es/. Accessed 22.5.10.
Aguiló, E., Alegre, J. and Sard, M. (2003) Examining the market structure of the German and UK tour operating industries through an analysis of package holiday prices. *Tourism Economics* 9 (3), 255–278.
Aguiló, E., Alegre, J. and Sard, M. (2005) The persistence of the sun and sand tourism model. *Tourism Management* 26 (2), 219–231.
Aguiló, E., Riera, A. and Rossello, J. (2005) The short term price effect of a tourist tax through the medium of a demand model: The case of the Balearic Islands. *Tourism Management* 26 (3), 359–365.
Aguiló, P., Alegre, J. and Riera, A. (2001) Determinants of the price of German tourist packages on the island of Mallorca. *Tourism Economics* 7 (1), 59–74.
Aitchison, C., Macleod, N. and Shaw, S. (2000) *Leisure and Tourism Landscapes.* London: Routledge.
Akhtar, M. and Humphries, S. (2000) *Some Liked It Hot: The British on Holiday at Home and Abroad.* London: Virgin Publishing Ltd.
Alegre, J. and Cladera, M. (2006) Repeat visitation in mature sun and sand holiday destinations. *Journal of Travel Research* 44 (3), 288–229.

Alenyar, M. (1990) Turisme i hostaleria. In *30 anys de turisme a Balears. Estudis Balearics* (Vol. 37–38, 17–37).

Alzina, J. *et al.* (1992) *Història de Mallorca* (2 vols). Palma: Moll.

Amelung, B. and Viner, D. (2006) Mediterranean tourism: Exploring the future with the Tourism Climatic Index. *Journal of Sustainable Tourism* 14 (4), 349–366.

Amer i Fernandez, J. (2006a) Empresaris hotelers i pacte de progres (1999–2003): Un enforntament mes enlla de ecotaxa (trans – a confrontation over ecotax). *Territoris* 6, 107–124.

Amer i Fernandez, J. (2006b) *Turisme i política: l'empresariat hoteler de Mallorca.* Palma: Documenta Balear.

Andreu, N. *et al.* (2003a) The 4th boom? Trends in natural resource consumption in the Balearic Islands. *Journal of Geography* 2, 61–77.

Andreu, N. *et al.* (2003b) *The Measurement of Sustainable Tourism in the Balearic Islands.* Palma: Center d'Investigació i Tecnologies de les Illes Balears tourist, CITTIB, Ministry of Tourism, Govern de les Illes Balears, Universitat de les Illes Balears.

Andrews, H. (2006) Consuming pleasures: Package tourists in Mallorca. In K. Meethan, A. Anderson and S. Miles (eds) *Tourism, Consumption and Representation: Narratives of Place and Self.* Wallingford: CABI.

Anon (2007) Mallorca: A growing number of corporate big guns are choosing the Balearic Island. *Conference and Incentive Travel* Sept., 77–80.

Aramberri, J. and Butler, R. (2004) *Tourism Development: Issues for a Vulnerable Industry.* Clevedon: Channel View Publications.

Arnau i Segarra, P. (1999) *Narrativa i turisme a Mallorca (1968–1980).* Palma. Documenta Balear.

Artigues, A. *et al.* (2006) *Introducción a la geografía urbana de las Illes Balears: VIII coloquio y jornadas de campo de geografia urbana, Illes Balears.* Palma: Govern de les Illes Balears and Conselleria de Turisme i Institut Balear del Turisme.

Ashworth, G. and Bruce, D. (2009) Town walls, walled towns and tourism: Paradoxes and paradigms. *Journal of Heritage Tourism* 4 (4), 199–213.

Baerenholdt, J., Haldrup, M., Larsen, J. and Urry, J. (2004) *Performing Tourist Places.* Aldershot: Ashgate.

Bahamonde Magro, A. (2005) Los límites de la modernización en España a principios de siglo XX. *Estudios turísticos* 163–164 (Issue dedicated to 100 years of Spanish tourist administration, 1905–2005), 7–16.

Baker, S. (2006) *Sustainable Development.* London: Routledge.

Ballester Vallori, A., Gelabert, M., Petrus Bey, J.M. and Gomila, R. (1993) *El Pla de Mallorca.* Palma: Alpha 3.

Barceló, A. (1961) El auge turístico de Mallorca. *Boletin de la camera official de comercio, indústria y navigacion de Palma de Mallorca (BOCIN)* 638, 138–141.

Barceló Pons, B. (1963) El Terreno. Geografica urbana de un barrio de Palma. *Boletin de la camera official de comercio, indústria y navigacion de Palma de Mallorca (BOCIN)* 640,125–178.

Barceló Pons, B. (1966) El turismo en Mallorca en la epoca de 1925–36. *Boletin de la camera official de comercio, indústria y navigacion de Palma de Mallorca (BOCIN)* 643, 651–662.

Barceló Pons, B. (1970) *Evolucion reciente y estructura actual de la populacion en las Islas Baleares.* Madrid: Instituto de Geografia Aplicada.

Barceló Pons, B. and Frontera Pascual, G. (2000) Historia del turismo en las Islas Baleares. In F. Tugores Truyol (coord.) *Welcome! Un siglo de turismo en las Islas Baleares* (pp. 15–36). Barcelona: Fundacion la Caixa.

Bardolet, E. (2001) The path towards sustainability in the Balearic Islands. In D. Ioannides, Y. Apostolopoulos and S. Sonmez (eds) *Mediterranean Islands and Sustainable Tourism Development: Practices Management and Policies* (pp. 193–213). London: Continuum.

Barke, M. (2007) Second homes in Spain: An analysis of change at the provincial level, 1981–2001. *Geography* 92 (3), 189–201.

Barke, M. and Towner, J. (1996) Exploring the history of leisure and tourism in Spain. In M. Barke, J. Towner and M. Newton (eds) *Tourism in Spain: Critical Issues* (pp. 1–34). Wallingford: CABI.

Barke, M., Towner, J. and Newton, M. (eds) (1996) *Tourism in Spain: Critical Issues.* Wallingford: CABI.

Bartholomew, E.G. (1869) Seven months in the Balearic Islands In H.W. Bates (ed.) *Illustrated Travels.* Vol 1. London. Cassell, Petter and Galpin.

Barton, S. (2004) Package holidays in Spain in the 1950s and 1960s. *Hospitality Review* 6 (3), 22–28.

Barton, S. (2005) *Working Class Organisations and Popular Tourism, 1840–1970.* Manchester: Manchester University Press.

Baum, T. (1998) Tourism marketing and the small island environment. In E. Laws, B. Faulkner and G. Moscado (eds) *Embracing and Managing Change in Tourism* (pp. 116–137). London: Routledge.

Baum, T. (2006) Revisiting the tourism area life cycle model – Is there an off-ramp? In R. Butler (ed.) *The Tourism Area Life Cycle, Vol. 2: Conceptual and Theoretical Issues* (pp. 219–233). Clevedon: Channel View Publications.

Beckett, B. (1947) *Memories of Mallorca.* London: Nelson.

Bennassar, B. (2001) *Proces al turisme. Turisme de masses, immigració, medi ambient i marginació a Mallorca (1960–2000).* Palma: Lleonard Muntaner.

Bidwell, C.T. (1876) *The Balearic Islands.* London. Samson, Low, Marston, Searle and Rivington.

Bigano, A., Goria, A., Hamilton, J. and Tol, R. (2005) The effect of climate change and extreme weather events on tourism. Working paper 30. Fondazione Eni Enrico Mattei. On WWW at www.feem.it. Accessed 6.6.10.

Binimelis, J. (1998) Les areés rururbanes a l'illa de Mallorca. *Estudis Balearics* 60–61, 185–203.

Binimelis, J. (2002) Canvi rural i colonització estrangera a Mallorca. In C. Picornell and A. Pomar (coords) *L'espai turístic. Planificació, gestió, recursos, sostenibilitat, noves modalitats* (pp. 207–236). Palma: INESE.

Blasco, A. (2002) Planificacion y gestion del territorio turistico de las islas Baleares. In D. Blanquer (ed.) *Ordenación y gestión del territorio turístico.* Valencia: Tirant lo Blanch.

Blázquez, M., Murray, I. and Garau, J. (2001) Indicadors de sostenibilitat del turisme i ordinacio del territori del les Illes Balears. In G. Xaviar Pons (ed.) *3rd Jornades de Medi Ambient de les Illes Balears* (pp. 42–51). Palma: Societe d'historic natural de les Illes Balears.

Blázquez Salom, M. (2005) El territorialismo y el ecologismo frente al turismo. *Scripta Nova. Revista electronica de geografia y sciences socials* IX, 194 (24).

Blázquez Salom, M. (2006) Calmar, contenir i decreixer. Polítques provades (1983–2003) i possibles de planificacio urbanistica. *Territoris* 6, 159–175.

Bows, A., Anderson, K. and Peeters, P. (2009) Air transport, climate change and tourism. *Tourism and Hospitality Planning & Development* 6 (1), 7–20.

Boyd, M.S. (1911) *The Fortunate Isles: Life and Travel in Majorca, Minorca and Ibiza.* London: Methuen and Co.

Bramwell, B. (2004) Mass tourism, diversification and sustainability in Southern Europe's coastal regions. In B. Bramwell (ed.) *Coastal Mass Tourism: Diversification and Sustainable Development in Southern Europe* (pp. 1–31). Clevedon: Channel View Publications.

Braudel, F. (1992) *The Mediterranean and the Mediterranean World in the Age of Philip II*. London: Collins Harper.

Bray, R. and Raitz, V. (2001) *Flight to the Sun: The Story of the Holiday Revolution*. London: Continuum.

Brendon, P. (1991) *Thomas Cook: 150 years of popular tourism*. London. Secker and Warburg.

Buades, J. (2004) *On brilla el sol. Turisme a Balears abans del boom*. Eivissa: Res Publica Edicion.

Buchanan, T. (2001) Receding triumph: British opposition to the Franco regime, 1945–59. *Twentieth Century British History* 12 (2), 163–184.

Buckley, R. (2002) Tourism ecolabels. *Annals of Tourism Research* 29 (1), 183–208.

Buckley, R. (2009) Large-scale links between tourism enterprises and sustainable development. In D. Leslie (ed.) *Tourism Enterprises and Sustainable Development: International Perspectives on Responses to the Sustainability Agenda*. London: Routledge.

Bull, P. (1997) Mass tourism in the Balearic Islands: An example of concentrated dependence. In D. Lockhart and D. Drakakis-Smith (eds) *Island Tourism: Trends and Prospects* (pp. 137–151). London: Pinter.

Burkhart, A. and Medlik, S. (1974) *Tourism - past, present and future*. London. Heinemann.

Buswell, R. (1996) Tourism in the Balearic Islands. In M. Barke, J. Towner and M. Newton (eds) *Tourism in Spain: Critical Issues* (pp. 309–340). Wallingford: CABI.

Buswell, R. (forthcoming) *Mallorca: The Making of the Landscape*.

Butler, R. (1998) Sustainable tourism – Looking backwards in order to progress. In C. Hall and A. Lew (eds) *Sustainable Tourism: A Geographical Perspective* (pp. 25–34). London: Addison Wesley Longman.

Butler, R. (ed.) (2006) *The Tourism Area Life Cycle, Vol. 1: Applications and Modifications*. Clevedon: Channel View Publications.

CAEB (2002) *Agroturismo y turismo rural en Balears*. Palma: CAEB and Govern Balear.

Cals, J. (1990) La política turística de l'Estat. In M. Alenyar (ed.) *30 anys de turisme a Balears*. *Estudis Baleàrics* (Vol. 37–38, 125–132).

Camps, M. (2004) Lo que traerá la ampliación del aeropuerto de Mallorca: un nuevo espaldarazo al turismo insostenible. *Quercus* 223, 80–81.

Cañellas Serrano, N. (1997) *El paisatge de l'Arxiduc*. Palma: Insitut d'estudis Balerics i Consellaria Medi Ambient, Ordinació de territori i litoral.

Capó Parrilla, J., Riera Font, A. and Rosselló Nadal, J. (2007) Accommodation determinants of seasonal patterns. *Annals of Tourism Research* 34 (2), 422–436.

Carbonero, M. and Salvà, P. (1989) Aspects geografics de la immigació a les Balears. In M. Alenyar i Fuster (ed.) *Les Migracions* (pp. 106–124). Palma: Ajuntament de Palma.

Carbonero Gamundi, M.A. (1991) Estructura rural i indústria a Palma, 1820–1930. In C. Manera and J.M. Petrus (eds) *Del taller a la fàbrica* (pp. 91–100). Palma: Ajuntament de Palma.

Caro Mesquida, M. (2002) El turisme a les Illes Balears (1905–1936). In A. Ordinas Garau and A. Ripoll Martínez (eds) *Turisme i societat a les Illes Balears*. Palma: Govern Balear and Conselleria de Turisme.

Carr, J. (1811) *Descriptive Travels in the Southern and Eastern Parts of Spain and the Balearic Islands in the year 1809*. London: Sherwood, Neely and Jones.

Carter, H. (1995) *The Study of Urban Geography* (4th edn). London: Arnold.

Cartier, C. (2005) Introduction: Touristed landscapes/seductions of place. In C. Cartier and A. Lew (eds) *Seduction of Place* (pp. 1–19). London: Routledge.

Cartwright, R. and Baird, C. (1999) *The Development and Growth of the Cruise Industry*. Oxford: Elsevier.

Casado-Diaz, M. (2004) Second homes in Spain. In C.M. Hall and D. Muller (eds) *Tourism, Mobility and Second Homes* (pp. 215–232). Clevedon: Channel View Publications.

Casanovas, M.-A. (2005) *L'economia balear (1898–1929)*. Palma: Documenta Balear.

Čavlek, N. (2000) The impact of tour operators on tourism development: A sequence of events. In J. Aramberri and R. Butler (eds) *Tourism Development: Issues for an Industry* (pp. 174–192). Clevedon: Channel View Publications.

Centre de Reçerca Economica UIB/Sa Nostra (2008) At www.cre.uib.es/.

Chamberlin, F. (1925) *Chamberlin's Guide to Majorca*. Barcelona: Augusta.

Cirer, J.C. (2006) *El turisme a les Balears (1900–1950)*. Palma: Documenta Balear.

Cirer, J.C. (2009) *La invenció del turisme de masses a Mallorca*. Palma: Documenta Balear.

Claver-Cortes, E., Molina Azorin, J.F. and Pereira Moliner, J. (2007) Competitiveness in mass tourism. *Annals of Tourism Research* 34 (3), 727–745.

Clayton, J. (1869) *The Sunny South: An Autumn in Spain and Majorca*. London: Hurst & Blackett.

Cobb, V. (2002) *The Package Tour Industry: The Pioneers and the Guinea Pigs*. London: Blackie and Co.

Coles, T. (2006) Enigma variations? The TALC marketing models and the descendents of the product life cycle. In R. Butler (ed.) *The Tourism Area Life Cycle, Vol. 2: Conceptual and Theoretical Issues* (pp. 49–66). Clevedon: Channel View Publications.

Coll Ramis, M., Feliu de la Peña Pons, J., Llabrés Prat, N., Romera Gallegos, P. and Rullan Bauzà, E. (2007) Evolución de la línea de costa de la Playa de Palma-Arenal (1956–2004). *Territoris* 7, 193–202.

Consell de Mallorca (2009) *Dades de la gestió dels residus urbans a Mallorca*. Dirrecció Insular de Gestió de Residus. On WWW at www.gestioderesidus.net.

Cormack, B. (1998) *A History of Holidays, 1812–1990*. London: Routledge/Thoemmes Press.

Crawford Flitch, J. (1911) *Mediterranean Moods*. London: Grant Richards.

Dades Informatives (various dates) *El turisme a les illes Balears*. Palma: Govern Balear, Conselleria de turisme and INESTUR-CITTIB.

Development of Rural Tourism in Mallorca: a strategy for sustainable management of the Balearic Islands? *Mediterranean Shores* [Online], 12, 2002. On WWW at http://rives.revues.org/137. Accessed 12.5.10.

Dillon, P. (2009) *Trekking through Mallorca – GR221 – the Drystone Route*. Milnethorpe: Cicerone.

Dodds, R. (2007) Sustainable tourism and policy implementation: Lessons from the case of Calvià, Spain. *Current Issues in Tourism* 10 (4), 296–322.

Dodds, R. (2008) *Power and Politics: sustainability in islands? Determining Barriers and Successes to Implementing Sustainable Tourism Policy in Two Mediterranean Islands: Calvià, Spain and Malta*. Berlin: VDM Verlag.

Dodds, R. and Kelman, I. (2008) How climate change is considered in sustainable tourism policies: A case of the Mediterranean islands of Malta and Mallorca. *Tourism Review International* 12 (1), 57–70.

d'Entrement (1993) El eje de desarrollo economico del Mediterráneo occidental. *Bulleti Balear d'Economia* 4, 37–50.

Elwell, C. (1995) *The Greater Aphrodisiad or Majorca*. Ferrington: London.

Escartin, J.M. (2001) *La ciutat amuntegada. Indústria del calcat, desenvolupament urbá i condicions de vida el la Palma contemporània (1840–1940)*. Palma: Documenta Balear.

Essex, S., Kent, M. and Newnham, R. (2004) Tourism development in Mallorca: Is water supply a constraint? *Journal of Sustainable Tourism* 12 (1), 4–28.

Estavan, A. (2001) La destruccion de Mallorca: vuelven las autopistas. In G. Xaviar Pons (ed.) *3rd Jornades de Medi Ambient de les illes Balears* (pp. 279–280). Palma: Societe de Historic Natural de les Illes Balears.

Estavan, A. and Llorente, N. (2001) La huella ecologica del turismo. Una primera aproximacion al caso de Mallorca. In *Societat i sostenibilitat, cicle de conferencies. Papers de medi ambient* (pp. 79–93). Palma: Sa Nostra.

Fernández Fúster, L. (1991) *Historia general del turismo de masas*. Madrid: Alianza.

Fiol Guiscafré, J. (1992) *Descobrint la mediterrània: viatgers anglesos per les Illes Balears i Pitiüses el segle XIX*. Palma: Miquel Font.

Florit, F. and Seguí, J.M. (1992) Le tourisme culturel comme possible régénérateur de l'image touristique d'un pays: le cas de Majorque. *Revue de Géographie de Lyon* 67 (1), 65–68.

Fornos, J. (1995) Enquadrement geologic, evolució estructural i sedimentologia de S'Albufera de Mallorca. In A. Martínez Taberna and J. Mayol Serra (eds) *S'Albufera de Mallorca*. Palma: Moll.

Forsyth, P., Hoque, T. *et al.* (2007) *The Carbon Footprint of Australian Tourism.* On WWW at www.crctourism.com.au. Accessed 22.5.10.

Fortuny, M., Soler, R., Cánovas, C. and Sánchez, A. (2008) Technical approach for a sustainable tourism development: Case study in the Balearic Islands. *Journal of Cleaner Production* 19 (7), 860–869.

Furlough, E. (1993) Packaging pleasures: Club Mediterranée and French consumer culture, 1950–68. *French Historical Studies* 18, 65–81.

Furlough, E. (2009) Club Mediterranée, 1950–2002. In I. Segreto, C. Manera, and M. Pohl (eds) *Europe at the Seaside: The Economic History of Mass Tourism in the Mediterranean* (pp. 174–195). New York: Berghahn Books.

Gago, A., Labandeira, X., Picos, F. and Rodriguez, M. (2009) Specific and general taxation of tourism activities: Evidence from Spain. *Tourism Management* 30 (3), 381–392.

Garau Taberner, J. and Manera, C. (2006) The recent evolution and impact of tourism in the Mediterranean: The case of island regions, 1990–2002. Working Paper 9. Fondazione Eni Enrico Mattei. On WWW at http://www.bepress.com/feem/paper9. Accessed 21.10.10

Garau Vadell, J. and Orfila-Sintes, F. (2008) Internet innovation for external relations in the Balearic hotel industry. *Journal of Business & Industrial Marketing* 23 (1), 70–80.

Garau Vadell, J. and de Borja-Sole, L. (2008) Golf in mass tourism destinations facing seasonality: A longitudinal study. *Tourism Review* 63 (2), 16–24.

Garau Vadell, J. and Serra Cantallops, A. (1998) *El turismo de golf en Balears*. Palma: Confederació d'Associacions Empresarials de Balears.

Garcia, C. and Servera, J. (2003) Impacts of tourism development on water demand and beach degradation on the island of Mallorca (Spain). *Geografiska Annaler: Series A, Physical Geography* 85 (3–4), 287–300.

Gines, A. and Garcia, L. (1995) El carst com a recurs turistic natural a Mallorca. In P. Salvà, A. Sastre and E. Aguilo (coords) *El desenvolupament turístic a la*

Mediterrània durant el segle XX. XIII Jornades d'estudis historics local (pp. 137–149). Palma: Institut d'estudis Balearics.

Giorgi, F. and Lionello, P. (2008) Climate change projections for the Mediterranean region. *Global and Planetary Change* 63 (2–3), 90–104.

Gladstone, D.L. (1998) Tourism urbanization in the United States. *Urban Affairs Review* 34 (1), 3–27.

Goldring, D. (1946) *Journeys in the Sun: Memories of Happy Days in France, Italy and the Balearic Islands.* London: Macdonald.

Gonzales Morales, J. (2005) La comision national de turismo y las primeras iniciatives para la fomento del turismo. La industria de los foresteros 1905–11. *Estudios Turísticos* 163–164, 17–30.

Gonzalez, M.A. (2001) Inmigracion y cohesion social en Calvià, Mallorca. *Nova Scripta. Revista electronica de geografia y sciences socials* 94, 21.

González Pérez, J. (2006) Geografía urbana de Palma: la actividad turística en la forma y el desarrollo de la ciudad. In *Introducción a la geografía urbana de las Illes Balears: VIII coloquio y Jornadas de campo de Geografia Urbana, Illes Balears*, 19–24 June, 164–210.

González Pérez, M. (2003) La pérdida de espacios de identidad y la construcción de lugares en el paisaje turístico de Mallorca. *Boletín de la Asociación de Geógrafos Españoles* 35, 137–152.

González Reverté, F. (2008) The role of tourist destinations in the social and demographic transformation of the Spanish Mediterranean coast. *Boletín de la Asociación de Geógrafos Españoles* 47, 371–373.

Govern Balear (1995) *P.O.O.T (Plan d'Ordinació de la Oferta Turística).* Palma: Conselleria de medi ambient i ordinació del territori.

Govern Balear (1997) *D.O.T: Analisis y diagnostico.* Palma: Conselleria de medi ambient, ordinació del territori i litoral.

Govern Balear, Ministry of Environment and Transport, Directorate General for Climate Change and Environmental Education (2008) Inventari d'emissions de contaminants a l'atmosfera de les Illes Balears.

Govern Balear (2010) *Nou sistema d'indicadors clau de sostenibilitat de les illes Balears (ICIB) per a les Agendes Locals 21 de les illes Balears.* Palma: Conselleria de medi ambient i mobilitat. On WWW at http://www.caib.es/sacmicrofront/archivopub. do?ctrl = MCRST96ZI77906&id = 77906. Accessed 10.6.10.

Grimalt Gelabert, M. (2002) Camps de golf, ports esportius, parcs aquatics i tematics. In A. Ordinas Garau and A. Ripoll Martínez (eds) *Turisme i societat a les Illes Balears* (pp. 289–305). Palma: Govern Balear and Conselleria de Turisme.

Gual, M., Moià, A. and March, J.G. (2008) Monitoring of an indoor pilot plant for osmosis rejection and grey water reuse to flush toilets in a hotel. *Desalination* 219 (1–3), 81–88.

Hall, C.M. and Higham, J. (eds) (2005) *Tourism, Recreation and Climate Change.* Cleveldon: Channel View Publications.

Hennessy, P. (2007) *Having It So Good: Britain in the Fifties.* London.

Holloway, C. (2006) *The Business of Tourism* (6th edn). London: Prentice-Hall.

Hughes, K., Bellis, M., *et al.* (2008) Predictors of violence in young tourists: A comparative study of British, German and Spanish holidaymakers. *European Journal of Public Health*, 18, 569–574.

ICCA Statistics Report (2008) *The International Association Meetings Market.* Amsterdam: International Congress and Convention Association.

INESTUR (2008) *Turismo urbano en la ciudad de Palma.* Palma: INESTUR.

INESTUR-CAEB (2005a) *Golf Tourism in the Balearic Islands.* Col.lecció estudis turístics no. 5. Palma: INESTUR.

INESTUR-CAEB (2005b) *Cruise Tourism in the Balearic Islands*. Col.leccció estudis turístics no. 7. Palma: INESTUR.

INESTUR-CAEB (2006) *El turisme de congressos a les Illes Balears*. Col.leccció estudis turístics no. 8. Palma: INESTUR.

INESTUR-CAEB (2007) *El turisme nàutic a Balears*. Col.leccció estudis turístics no. 9. Palma: INESTUR.

Institut Balear de Turisme (2001) *Mallorca 20th Century: An Essential Destination*. Palma: Diario de Mallorca.

Instituto de Estudios Turísticos (2009) *Compañías aereas de bajo coste*. On WWW at www.iet.tourspain.es/informesdocumentacion/cbc/bajo coste 2009.pdf. Accessed 15.6.10.

Ioannides, D. (1998) Tour operators: The gatekeepers of tourism. In D. Ioannides and K. Debbage (eds) *The Economic Geography of the Tourism Industry: A Supply Side Analysis* (pp. 139–158). London: Routledge.

IRN Research (2009) Annual cruise review. On WWW at www.irn-research.com. Accessed 15.5.10.

King, R. (1993) The geographical fascination of islands. In D.G. Lockhart, D. Drakakis-Smith and J. Schembri (eds) *The Development Process in Small Island States* (pp. 13–37). London: Routledge.

Knowles, T. and Curtis, S. (1999) The market viability of European mass tourist destinations: A post-stagnation life cycle analysis. *International Journal of Tourism Research* 1, 87–96.

Kozak, M. (2001) Comparative assessment of tourist satisfaction with destinations across two nationalities. *Tourism Management* 22 (4), 391–401.

Kozak, M. (2002) Comparative analysis of tourist motivations by nationality and destinations. *Tourism Management* 23, 221–232.

Kozak, M. and Rimmington, M. (2000) Tourist satisfaction with Mallorca, Spain, as an off-season holiday destination. *Journal of Travel Research* 38 (3), 260–269.

Kynaston, D. (2007) *Austerity Britain, 1945–51*. London: Bloomsbury.

Lawrence, D.H. (1955) The man who loved islands. In *Collected Short Stories* (Vol. 3). London: Heinemann.

L'aéroport de Palma, clé du tourisme à Majorque. *Rives méditerranéennes* [on line] 12, 2002. On WWW at http://rives.revues.org/135. Accessed 12.6.10.

Lenček, L. and Bosker, G. (1999) *The beach; the history of paradise on earth*. London. Pimlico.

LIRC (2009) *Leisure Forecasts 2009–2013*. London: LIRC.

Llado, M. (2003) Turismo residencial y dispersion urbana en Mallorca (Illes Balears). Un ensayo metodologico en el municipio de Pollensa. *Estudios Turísticos* 155/6, 197–218.

Llibre Blanc (1987) *Llibre blanc del turisme de les Illes Balears*. Palma: Govern Balear, UIB *et al.*

Llibre Blanc (2009) *Llibre blanc del turisme de les Illes Balears: capa una nova cultura turística*. Palma: Govern Balear, UIB *el al.*

Llul Gilet, A. (coord.) (2002) *La empresa turistica balear y el medio ambiente*. Palma: UIB.

Lohmann, M. (2001) Coastal resorts and climate change. In A. Lockwood and S. Medlik (eds) *Tourism and Hospitality in the 21st Century* (pp. 284–295). Oxford: Butterworth-Heinemann.

Lohmann, M. and Kaim, E. (1999) Weather and holiday destination preferences, image attitude and experience. *The Tourist Review* 2, 54–64.

López Bravo, P. (2003) Impacts of tourism on Majorca: The new colonisation. Unpublished MA thesis, Bournmouth University.

López Nadal, G. and Maturana Bis, F. (2008) Fenicios en una renovada carrera de Indias. Paper presented to XV Congreso Internacional de Ahila. Crisis y problemas en el mundo Atlántico, 1808–2008, 26–29 August 2008.

Lyth, P. (2003) Gimme a ticket on an aeroplane: The jet engine and the revolution in leisure air travel c.1960–1990. In L. Tissot (ed.) *Construction of a Tourism industry in the Nineteenth and Twentieth Centuries.* Neuchatel, Switzerland: Alphil.

Lyth, P. (2009) Flying visits:the growth of British air package tours In, Segreto, L. Manera, C. and Pohl, M. (Eds) *Europe at the seaside; the economic history of mass tourism in the Mediterranean.* Oxford. Berghahn Books, 11–30.

Lyth, P. and Dierikx, M.L.J. (1994) From privilege to popularity: The growth of leisure air travel since 1945. *Journal of Transport History* 15 (2), 97–116.

Mallorca Daily Bulletin (4 June 2008) First phase of the tram will run to airport.

Mallorca Daily Bulletin (6 June 2008) Superliner makes her maiden visit.

Mallorca Daily Bulletin (10 June 2008) The Sunday interview; Jonathan Syrett, Vice-President Spanish Large Yacht Association.

Mallorca Daily Bulletin (21 October 2009) 'If the demand was there, airlines would be flying' explains airline boss.

Mallorca Daily Bulletin (27 October 2009) An extra 1200 moorings created in two years.

Mallorca Daily Bulletin (29 October 2009) Playa de Palma reform 'master plan' to be ready in May next year.

Mallorca Daily Bulletin (30 October 2009) Tourist spending slumps in the first nine months of 2009.

Mallorca Daily Bulletin (14 April 2010) Conferences.

Malvárez, G. Pollard, J. and Domínguez, R. The planning and practice of coastal zone management in southern Spain. In B. Bramwell (ed.) *Coastal Mass Tourism: Diversification and Sustainable Development in Southern Europe* (pp. 200–219). Clevedon: Channel View Publications.

Manera, C. (2001) *Història del creixement econòmic a Mallorca, 1700–2000.* Palma: Lleonard Muntaner.

Manera, C. (2006) *La riqueza de Mallorca; una història econòmic.* Palma: Lleonard Muntaner.

Manera, C. (2010) *La recta raó; economia, història econòmica i sostenibilitat a les Illes Balears.* Palma: Moll.

Manera, C. and Garau-Taberner, J. (2009) The transformation of the economic model of the Balearic Islands: The pioneers of mass tourism. In I. Segreto, C. Manera and M. Pohl (eds) *Europe at the Seaside: The Economic History of Mass Tourism in the Mediterranean* (pp. 31–48). New York: Berghahn Books.

Manera, C. and Petrus Bey, M.A. (1991) El sector indùstrial en el creixment econòmic de Mallorca 1780–1985. In C. Manera and M.A. Petrus Bey (eds) *Del taller a la fàbrica* (pp. 13–58). Palma: Ajuntament de Palma.

Marcos, M. and Tsimplis, M. (2008) Comparison of results of AOGCMs in the Mediterranean Sea during the 21st century. *Journal of Geophysical Research* 113, C12028, doi:10.1029/2008JC004820. Accessed 7.8.10.

Martinez Taberna, A. and Mayol Serra, J. (eds) (1995) *S'Albufera de Mallorca.* Palma: Moll.

Martin-Prieto, J., Roig-Munar, F. and Rodriguez-Perea, A. (2007) Analisis espacio-temporal (1956–2005) de la foredune de Cala Mequida (N. Mallorca). Mediante eo uso de variables geoambientales y autotropicas. *Territoris* 7, 175–191.

Mascaró Pons, P. (1989) Migracions i mercat de treball. In M. Alenyar i Fuster (ed.) *Les Migracions*. Palma: Ajuntament de Palma, (pp. 93–104).

Mata Pastor, J. (2002) El turisme a les Illes Balears (1973–83). In A. Ordinas Garau and A. Ripoll Martínez (eds) *Turisme i societat a les Illes Balears*. Palma: Govern Balear and Conselleria de Turisme, 81–95.

Mayol, J. and Machado, A. (1992) *Medi ambient, ecologia i turisme a les illes Balears*. Palma: Moll.

Middleton, V. (1991) Whither the package tour? *Tourism Management* 12 (3), 185–192.

Miranda Gonzalez, M. (2001) Immigration and social cohesion in Calvià, Mallorca. Proceedings of the Third International Symposium on migration and social change. *Scripta Nova. Electronic Journal of Geography and Social Sciences* 94 (21), 1.

Ministerio de Econòmia (2006) *Turismo de reuniones*. Estudios de productos turisticos no. 5. Secretaria general de turismo. Madrid: TURESPAÑA.

Ministerio de informacion y turismo 1965: Estatísticas de turismo. Madrid. Ministerio de informacion y turismo.

Mintel Travel and Tourism Analyst (2007) Report no.1: Climate change.

Miralles, S. (2001) El transport com un factor limitador del creixement. In G. Xaviar Pons (ed.) *3rd Jornades de Medi Ambient de les illes Balears* (pp. 277–278). Palma: Societé d'historic natural de les Illes Balears.

Monfort Mir, V. and Ivars Baidel, J. (2001) Towards a sustained competitiveness of Spanish tourism industry. In Y. Apostolopolous, P. Loukissas and L. Leontidu (eds) *Mediterranean Tourism: Facets of Socioeconomic Development and Cultural Change* (pp. 17–38). London: Routledge.

Monserrat i Moll, A. (1990) El turismo y el mercado de trabajo en Baleares. *Estudis Baleàrics* 37/38, 97–107.

Montanari, A. (1995) The Mediterranean region: Europe's summer leisure space. In A. Montanari and A.M. Williams (eds) *European Tourism: Regions, Spaces and Restructuring* (pp. 41–65). Chichester: Wiley.

Moreno Rodriguez, F. (1990) La ensenanza profesional turistica en Baleares (1960–1989). *Estudis Baleàrics* 37/38, 109–124.

Morgan, M. (1991) Dressing up to survive: Marketing Majorca anew. *Tourism Management* 12 (1), 15–20.

Mulet, A. (1945) *Importancia de turismo en Mallorca*. Extracted from BOCOCIN nos. 563, 564 and 565 and published separately. Palma: BOCOCIN.

Mullins, P. (1991) Tourism urbanization. *International Journal of Urban and Regional Research* 15 (3), 326–342.

Mullins, P.F. (2003) Cities for pleasure – The emergence of tourism urbanization in Australia. In D-Y. Jeong and P. Mullins (eds) *The Environment and Sustainable Development* (pp. 234–251). South Korea: Cheju National University Press.

Murray, I., Rullan, O. and Blázquez, M. (2005) Traces of territorial ecological deterioration: Background to the occult explosion tourism in islands. *Scripta Nova. Electronic Journal of Geography and Social Sciences* IX, 199.

Nájera, M. and Bustamante, J. (2007) La experiencia de las Agendas Locales 21 en destinos turísticos: el caso de Calvià (Baleares). *Estudios Turisticos* 172–173, 97–106.

Newton, M. (1996) Tourism and public administration in Spain. In M. Barke, J. Towner and M. Newton (eds) *Tourism in Spain: Critical Issues* (pp. 137–166). Wallingford: CABI.

O'Hare, D. (1997) Interpreting the cultural landscape for tourist development. *Urban Design* 2 (1), 33–54.

Ordinas Garau, A. (2000) El cami de vell de Lluc: un exemple de les potencialitats del serendisme com a modalitat turistica a la Serra de Tramuntana de Mallorca. In M. Ramos Moray (ed.) *Evolució turistica de la derrera decada i disseny de futur. Actes de les Jornades de Turisme i Medi Ambient a les Illes Balears* (pp. 399–404).

Ordinas Garau, A. and Jaume Binimelis Sebastián, J. (2003) El turismo de reuniones y negocios en Mallorca. *Cuadernos de turismo* 12, 35–52.

O'Reilly, A. (1986) Tourism carrying capacity: Concepts and issues. *Tourism Management* 7, 254–258.

Oreja Rodriguez, J., Parra-Lopez, E. and Yanes-Estevez, V. (2008) The sustainability of island destinations: Tourism area life cycle and teleological perspectives. The case of Tenerife. *Tourism Management* 29 (1), 53–65.

Orfila-Sintes, F., Crespí-Cladera, R. and Martínez-Ros, E. (2005) Innovation activity in the hotel industry: Evidence from the Balearic Islands. *Tourism Management* 26 (6), 851–865.

Orfila-Sintes, F. and Mattsson, J. (2009) Innovation behavior in the hotel industry. *Omega* 37 (2), 380–394.

Pack, S. (2006a) *Tourism and Dictatorship: Europe's Peaceful Invasion of Franco's Spain.* New York: Palgrave.

Pack, S. (2006b) Tourism, modernization and difference: A twentieth century Spanish paradigm. *Sport and Society* 11 (6), 657–672.

Pack, S. (2007) Tourism and political change in Franco's Spain. In N. Townson (ed.) *Spain Transformed: The Late Franco Dictatorship, 1959–75* (pp. 47–66). Basingstoke: Palgrave Macmillan.

Page, S. (1995) *Urban Tourism.* London: Routledge.

Palmer, C. (2004) More than just a game: The consequences of golf tourism. In B. Ritchie and D. Adair (eds) *Sport Tourism: Interrelationship, Impacts and Issues.* Clevedon: Channel View Publications.

Palmer, T. and Riera, A. (2003) Tourism and environmental taxes with reference to the Balearic ecotax. *Tourism Management* 24 (6), 665–674.

Palmer, T., Riera-Font, A. and Rosselló-Nadal, J. (2007) Taxing tourism: The case of rental cars in Mallorca. *Tourism Management* 28 (1), 271–279.

Pellejero Martínez, C. and Martin Rojo, I. (1998) Origen, desarrollo y consolidacion de un lider hotelero: Sól Melia, 1956–1997. *Tourism Review* 53 (2), 48–54.

Perry, A. (2000) Impacts of climate change on tourism in the Mediterranean: Adaptive responses. Working Paper 35. Fondazione Eni Enrico Mattei. On WWW at www.feem.it. Accessed 5.4.10.

Perry, A. (2004) The Mediterranean: How can the world's most popular and successful tourist destination adapt to a changing climate? In C.M. Hall and J. Higham (eds) *Tourism, Recreation and Climate Change* (pp. 86–96). Clevedon: Channel View Publications.

Picornell, C. (1986) *Excursion a la Playa de Palma.* 25–30 August. Mimeo. Palma: IGU Commission of Geography of Tourism and Leisure.

Picornell, C. (1990) Turisme i territorí a les Illes Balears; 18 conclusions generals sobre la geografia, la história, els impacts i la politica del turisme a les illes Balears. *Treballs de Geografia* 43, 43–48.

Picornell, C. and Picornell, M. (2002a) L'espai turistic de les illes Balears. Un cicle de vida d'una area turistíc? Evolucio i planificació a la derrera decada. In C. Picornell and A. Pomar (eds) *L'espai turistic* (pp. 31–96). Palma: INESA-GITTO.

Picornell, C. and Picornell, M. (2002b) El turisme a les Illes Balears (1983–2002). In A. Ordinas and A. Ripoll (eds) *Turisme i societat a les Illes Balears.* Palma: Govern Balear and Conselleria de Turisme.

Picornell, C. and Picornell, C. (2008) *Apunts del Pla de Mallorca*. Pollença: El Gall/ Institut d'Estudis Baleàrics.

Picornell, C. and Segui, J.M. (1989) *Geografía humana de las Isles Baleares*. Barcelona: Oikos-Tau.

Pons, A. (2004) Dels evolució uses de sol a Isles Balears, 1956–2000. *Territoris* 4, 129–145.

Pontin, F. (1991) *My Life. Always ... Thumbs Up*. London: Solo Books Ltd.

Pujalte Vilanova, F. (2002) *Transports i comunicaions a les Illes Balears durant el segle XX*. Quaderns d'Historia Contemporania de les Balears, No. 35. Palma: Documenta Balear.

Pujalte Vilanova, F. (2006) Electricitat i el procés d'industrialització a Mallorca: algunes reflexions. In F. Pujalte Vilanova (ed.) *Un segle de llum a Inca, 1905–2005*. Palma: Documenta Balear.

Quintana Peñuela, A. (1979) *El sistema urbano de Mallorca*. Palma: Moll.

Ringer, G. (ed.) (1998) *Destinations: Cultural Landscapes of Tourism*. London: Routledge.

Roca, J. (2006) La indùstriá a Mallorca. Quaderns d'Historia Contemporania de les Baleares, No. 48. Palma: Documenta Balear.

Roca, J. and Umbert, J. (1990) Economia y desarrollo industrial en Mallorca (1914–30), Apuntes de investigcion. *Estudios de Historia Economica* 1, 93–112.

Rodriguez Perea, A. (2010) Water resources, biodiversity and management of territory. Course organized by the Ministry of Justice and the Center for Legal Studies, CAIB, 19 April.

Rodriguez Perea, A., Servera Nicolau, J. and Martin Prieto, J.A. (2000) *Alternatives a la dependència de les platges de les Balears de la regeneració artificial continuada: informe Metadona*. Pedagogia Ambiental, No. 10. Palma: UIB.

Roig i Minar, F., Rodriguez Perca, T. and Martin Prieto, J. (2006) Analysis critico de las medidas de valoraçion en la calidad turística y ambiental de los sistemas litorales arenosas. *Territoris* 6, 27–44.

Royle, S. (1989) A human geography of islands. *Geography* 74,106–116.

Ruijgrok, G. and van Paassen, D. (2005) *Elements of Aircraft Pollution*. Amsterdam: Delft University Press.

Ruiz-Viñals, C. (2000) *L'urbanisme de la ciutat de Palma de Mallorca*. Palma: El Far de les Crestes.

Rullan Salamanca, O. (1989) El comportamant municipal de l'oferta de places turistiques a Mallorca entre 1965 i 1985. *Treballs de geografia* 41, 99–105.

Rullan Salamanca, O. (2002) *La construcció territorial de Mallorca*. Palma. Moll.

Rullan Salamanca, O. (2005) Planning technique to contain the residential growth areas with high pressure properties. *Scripta Nova. Electronic Journal of Geography and Social Sciences* IX, 194, 32.

Rullan Salamanca, O. (2006) Visita a la zona de ocio nocturno de la playa de Palma (Les Meravelles). In *Introducción a la geografía urbana de las Illes Balears: VIII coloquio y Jornadas de campo de Geografia Urbana* (pp. 163ff). Palma: Govern de les Illes Balears.

Rullan Salamanca, O. (2007a) L'ordinació territorial a les illes Balears (Segles XIX– XX). Quaderns d'Historia Contemporania de les Illes Ballears, No. 53. Palma: Documenta Balear.

Rullan Salamanca, O. (2007b) Isolated buildings or residence? Unique regions or areas? Boom or development? Rural or rustic space? *Scripta Nova. Revista Electrónic Journal of Geography and Social Sciences* XI, 232.

Russell, R. (2006) The contribution of entrepreneurship theory to the TALC model. In R. Butler (ed.) *The Tourism Area Life Cycle, Vol. 2: Conceptual and Theoretical Issues* (pp. 105–123). Clevedon: Channel View Publications.

Rydin, Y. (1993) *The British Planning System*. London: Macmillan.

Salas Colom, A. (1992) *El turismo en Mallorca; 50 años de historia*. Palma: Graficas Planisi.

Salvà Tomàs, P. (1986) Caracteristiques fonamental de l'home a les Balears: aspectes geografics de la populació de les illes. *Revista del Centre d'Estudis Teologics de Mallorca*, Nov–Dec, 3–36.

Salvà Tomàs, P. (1990) El turisme com a element impulsor del process d'urbanització a Balears (1960–1989). In M. Alenyar Fuster (coord.) *30 anys de turisme a Balears. Estudis Baleàrics* (37–38, 63–70) Palma.

Salvà Tomàs, P. (1992) Els effects de la transicio demografica illença sobre el territorí: el marc de l'emigració a les illes Balears entre 1878 i 1955.*Papers to Congres Internacional d'estudis Historics* (pp. 405–411). Palma.

Salvà Tomàs, P. (2000) El turisme de mañana. In F. Tugores i Truyol (coord.) *Welcome! Un siglo del turismo en las Islas Balears* (pp. 124–134). Barcelona: 'La Caixa'.

Salvà Tomàs, P. (2002a) Foreign immigration and tourism development in Spain's Balearic Islands. In C. Hall and A. Williams (eds) *Tourism and Migration: New Relationships between Production and Consumption* (Chap. 6). Berlin: Springer.

Salvà Tomàs, P. (2002b) Tourist development and foreign immigration in Balearic Islands. *Revue européenne des migrations internationals* 18 (1), 87–101.

Salvà Tomàs, P. and Binimelis Sebastian, J. (1993) Las residençias secundaris en la isla de Mallorca: tipos y procesos de crecimiento. *Mediterranée* 77 (1–2), 73–76.

Sand, G. (1956) *Un hiver en Majorque*. Winter in Majorca Valldemossa. (R. Graves, trans.) (original work published 1855).

Sastre, A. (1995) *Mercat turistíc Balear*. Palma: Institut d'Estudis Baleàrics.

Sastre, F. and Benito, I. (2001) The role of transnational tour operators in the development of Mediterranean island tourism. In D. Ioannides, Y. Apostolopoulos and S. Sonmez (eds) *Mediterranean Islands and Sustainable Tourism Development: Practices, Management and Policies* (pp. 69–86). London: Continuum.

Sbert Barceló, T. (2007) *Una evolucion turistica: historia de la Playa de Palma 1900–2006* (2nd edn). Palma: Asociacion empresarial de actividades turisticas de la Playa de Palma.

Seguí Aznar, M. (1999) L'arquitectura i l'urbanisme en el canvi de decada (1920–30). El programa municipal de l'Assembla dels partits Autonomista i Regionalista. In A. Marimar Riutort and S. Serra Busquets (eds) *Els anys vint a les illes balears*. 17th *Jornades d'estudis historícs locals* (pp. 57–68). Palma: Institut d'Estudis Balearics.

Seguí Aznar, M. (2001) *La arquitectura del ocio en Baleares*. Palma: Lleonard Muntaner.

Seguí Llínas, M. (1995) *Les novelles baléares. La renovation d'un espace mythique*. Paris. Ed L'Harmattan

Seguí Llínas, M. (2006) *El turisme a les Balears (1950–2005)*. Quaderns d'Historia Contemporania de les Balears, No. 51. Palma: Documenta Balear.

Seguí Pons, J. and Martinez Reynés, M. (2002) Saturació de la demanda i mesures de sostenibilitat a les Illes Balears. L'horitzo 2000. In C. Picornell Bauza and A. Pomar Goma (eds) *L'espai turistic* (pp. 259–282). Palma: INESA-GITTO.

Seguí Pons, J., Martínez Reynés, M. and Ruiz Pérez, M. (2007) Ruido y sostenibilidad ambiental en el aeropuerto de Son San Joan (Mallorca) *Cuadernos de geografía* 81–82, 51–70.

Selwyn, T. (2001) Tourism, development and society in the insular Mediterranean. In D. Ioanniddes, Y. Apostolopoulos and S. Sonmez (eds) *Mediterranean Islands and Sustainable Tourism Development: Practices, Management and Policies* (pp. 23–44). London: Continuum.

Serra, A. (2009) The expansion strategies of the Majorcan hotel chains. In I. Segreto, C. Manera and M. Pohl (eds) *Europe at the Seaside: The Economic History of Mass Tourism in the Mediterranean* (pp. 125–143). New York: Berghahn Books.

Serra i Busquets, S. (2000) La platje de Palma entre el present i el future. In M.A. Ramos Moray (ed.) *Evolució turistica de la derrera decada i disseny de futur. Actes de les Jornades de Turisme i Medi Ambient a les Illes Balears*. 81–88.

Serra i Busquets, S. and Company i Mates, A. (2000) El turismo en las instituciones y en el debate publico. In F. Tugores i Truyol (coord.) *Welcome! Un siglo del turismo en las Islas Balears* (pp. 70–86 (Spanish), 161–167 (English)). Barcelona: 'La Caixa'.

Servera Nicolau, J. (2004) *Geomorphologia del litoral de les Illes Balears*. Palma: Documenta Balear.

Shaw, G. and Williams, A. (1994) *Critical Issues in Tourism: A Geographical Perspective*, Oxford. Blackwell.

Shaw, G. and Williams, A. (2004a) *Critical Issues in Tourism: A Geographical Perspective* (2nd edn). London: Blackwell.

Shaw, G. and Williams, A. (2004b) *Tourism and Tourism Spaces*. London: Sage.

Shor, J. and Shor, F (1957) The Balearics are booming: Low-cost living beneath azure skies lures vacationing throngs to Spain's sunny Mediterranean 'Isle of Peace' (Majorca). *National Geographic Magazine* 111 (5), 621–660.

Sinclair, M. and Bote Gomez, V. (1996) Tourism: The Spanish economy and the balance of payments. In M. Barke, J. Towner and M. Newton (eds) *Tourism in Spain: Critical Issues* (pp. 89–117). Wallingford: CABI.

Social and Economic Report of Balearic Islands (various dates) Annual reports of El Centre de Recerca Econòmica (CRE), University of the Balearic Islands (UIB)-Sa Nostra, 2001 to date. On WWW at www.cre.uib.es. Accessed 8.11.10.

Spilanis, I. and Vayanni, H. (2004) Sustainable tourism: Utopia or necessity? The role of new forms of tourism in the Aegean Islands. In B. Bramwell (ed.) *Coastal Mass Tourism: Diversification and Sustainable Development in Southern Europe* (pp. 269–291). Clevedon: Channel View Publications.

Standeven, J. and De Knop, T. (1999) *Sport Tourism*. Champaign, IL: Human Kinetics.

Sumner, G., Ramis, C. and Guijarro, J. (1993) The spatial organization of daily rainfall over Mallorca, Spain. *International Journal of Climatology* 13 (1), 89–109.

Taylor, B. (1867–1868) By-ways of Europe I and II: A visit to the Balearic Islands. *Atlantic Monthly* 20 (122), 680–695 and 21 (123), 73–87.

Thomas, D. (2005) *Villains' Paradise: Britain's Underworld from the Spivs to the Krays*. London: John Murray.

Towner, J. (1996) *An Historical Geography of Recreation and Tourism in the Western World, 1540–1940*. Chichester: Wiley.

Trauer, B. and Ryan, C. (2005) Destination image, romance and place experience – An application of intimacy theory in tourism. *Tourism Management* 26 (4), 481–491.

Travel Foundation (2007) Tourism destinations carbon footprints: A report prepared by Dick Sisman & Associates. On WWW at http://www.teamfortheworld.org/docs/articles/tourismdestinations.pdf. Accessed 21.5.10.

Trelford, P. (2006) No job too big – or too small – New facilities are opening Mallorca up for large groups, but UK buyers are also realising the island's great potential as a top incentive travel destination. *Meetings and Incentive Travel* Jan, 41–48.

Tsimplis, M., Marcos, M. and Somot, S. (2008) 21st Century Mediterranean sea level rise: Steric and atmospheric pressure contributions from a regional model. *Global and Planetary Change* 63 (2–3), 105–111.

Tugores i Truyol, F. (coord.) (2000) *Welcome! Un siglo del turismo en las Islas Balears.* Barcelona: 'La Caixa'.

Urry, J. (2002) *The Tourist Gaze* (2nd edn). London: Sage.

Valenzuela, M. (1988) Spain: The phenomenon of mass tourism. In A. Williams and G. Shaw (eds) *Tourism and Economic Development* (pp. 39–57). London: Belhaven.

Vera Rebollo, J. and Ivars Baidal, J. (2008) Spread of low-cost carriers: Tourism and regional policy effects in Spain. *Regional Studies* 43 (4), 559–570.

Vich i Martorell, G. and Pou, L. (2007) The use of the Internet in the hotel sector of the Balearic Islands: Evolution and perceptions. In A. Matias, P. Nijkamp and P. Neto (eds) *Advances in Modern Tourism Research* (pp. 307–324). Heidelberg: Physica-Verlag.

Viner, J. and Agnew, M. (1999) *Climate Change and Its Impacts on Tourism* (report). Godalming, UK: WWF-UK.

WWF-UK (2002) *Holiday Footprinting: A Practical Tool for Responsible Tourism.* On WWW at http://www.wwf.org.uk/filelibrary/pdf/holidayfootprintingfull. pdf. Accessed 21.5.10.

WTO (1998) *Tourism Taxation* (The Watson report). Geneva: UNWTO.

WTO (2003) *Proceedings of the First International Conference on Climate Change and Tourism,* Djerba, Tunisia, 9–11 April.

WTO (2008) At http://:www.unwto.org/sustainable/doc/davos_sum/Climate-Change_Summary.pdf. Accessed 20.5.10.

Waldren, J. (1996) *Insiders and Outsiders: Paradise and Reality in Mallorca.* Oxford: Berg.

Waldren, J. (1997) We are not tourists, we live here! In S. Abram, J. Waldren and D. MacLeod (eds) *Tourists and Tourism: Identifying with People and Places* (pp. 51–70). Oxford: Berg.

Walton, J. and Walvin, J. (1983) *Leisure in Britain, 1780–1939,* Manchester: Manchester University Press.

Walton, J. (1994) British perceptions of Spain and their impact on attitudes to the Spanish Civil War: Some additional evidence. *Twentieth Century British History* 5, 283–299.

Walton, J. (2002) British tourism between industrialisation and globalization: An overview. In H. Berghoff, B. Korte, R. Schneider and C. Harvie (eds) *The Making of Modern Tourism: The Cultural History of the British Experience, 1600–2000* (pp. 109–131). Basingstoke: Palgrave.

Walton, J. (2005) Paradise lost and found: Tourists and expatriates in El Terreno, Palma de Mallorca, from the 1920s to the 1950s. In J. Walton (ed.) *Histories of Tourism: Representation, Identity, and Conflict* (pp. 179–194). Clevedon: Channel View Publications.

Walton, J. (2005a) Review of Susan Barton: *Working-Class Organisations and Popular Tourism, 1840–1970.* Reviews in History, 458. Manchester: Manchester University Press. On WWW at http://www.history.ac.uk/reviews/review/458. Accessed 11.6.09.

Walton, J. (2009) Prospects in tourism history: Evolution, state of play and future developments. *Tourism Management* 30, 783–793.

Ward, C. and Hardy, D. (1986) *Good-night, Campers!: History of the British Holiday Camp*. Series No. 9: Studies in History Planning & the Environment Series. London: Mansell.

Waring, H. (1877–1878) The drainage, irrigation and cultivation of the Albufera of Alcudia, Majorca, and application of the common reed as a material for paper. *Minutes of the Proceedings of the Institution of Civil Engineers* 52 (Part II), 243–249.

Waxman, P. (1933) *What Price Mallorca*. New York: Farrar & Rinhart.

Williams, A. and Shaw, G. (1998) Tourism and environment: Sustainability and economic restructuring. In C. Hall and A. Lew (eds) *Sustainable Tourism: A Geographical Perspective* (pp. 49–59). London: Addison Wesley Longman.

Wineaster, A., Juaneda, C. and Sastre, F. (2009) Influences of pro-all-inclusive travel decisions. *Tourism Review* 64 (2), 4–18.

Wood, C. (1888) *Letters from Mallorca*. London: Richard Bentley & Son.

Wright, S. (2002) Sun, sea, sand and self-expression. In H. Berghoff, B. Korte, R. Schneider and C. Harvie (eds) *The Making of Modern Tourism: The Cultural History of the British Experience, 1600–2000* (pp. 181–202). Basingstoke: Palgrave.

Index

A

ABTA scheme, 80
Aegean islands, 182
Africa, 91–92, 98, 155–156
agriculture, 39
    and the economy, 32–35
    industry and, 32–33, 35
    landscape and, 33–34
    rationalisation of, 68
agrotourism, 27, 93, 159–160, 165, 166
aircraft design and performance, 62
airlines, 48, 93
    charter, 63–64
airports, 63
Albufera, 3, 28–29, 40, 127
    ecological importance, 29
    lagoon-like feature, 28
Alcúdia, 90
alga, protection of, 20, 22
Algeria, 6–7
Alzamora family, role, 39–40
Andalucía, 155
Andratx, 16, 40
Aquarium Platje de Palma, 180
Artá caves, 33
Asia, 155–156, 159
Atlantic coasts, 10
Austro-Hungarian emperors, 4
autonomia, 125

B

Balearic Islands, 6–7, 15, 38, 60, 79, 85, 88,
    91–93, 97–98, 103, 106, 108–109, 112–114,
    118, 120–122, 127, 131–135, 140, 144, 147,
    151, 154–155, 157, 159
Balearic model, 10, 31, 145, 148, 150, 153
    history of, 178–179
Baltic States, 1, 10, 150
banking, advances in, 37
Banyabulfar's terraces, 33
Barceló group, 147

Barcelona, 154
bays, 8
beach culture, 44
beaches, 8, 18–25
    beach-dune dynamics, 18, 20, 23
    damage from storms, 20–22
    destruction and erosion relationship,
        20, 23
    management problems, 23
    Mediterranean climate and, 17
    natural forms and processes, 22, 25
    natural process of formation, 20
    as resource, 17
    in 1940s, 17–18
    in 1950s, 18
    tourists' interest in, 11–12
    types of, 18–19
berths, 68, 113, 167–169
bioclastic material, 20
Blackpool, 1
Britain, 52, 53, 78, 80, 80–81, 120, 174
British travelers, 2–3, 5, 52–53
Buades, Joan, 50–51
Butler cycle, 148
Butler model, 31–32, 81, 82, 176, 180
Butler, Richard, 101–102, 148, 176

C

Cabrera island, 15
Cala Bona, 105
Cala d'Or, 18–19, 44–45
Cala Figuera, 18
Cala Mesquida, 23
Cala Millor, 65, 105, 186
Cala Murada, 18
Calas de Mallorca beach, 21, 24
Calviá, 16, 46, 83, 86–87, 90, 102–103, 105,
    109, 128, 139, 168
Campos, 90
car hiring, 90, 104, 112, 113, 118, 179
Caribbean, 145, 156–157
Cas Català, 56